THE EXPLORATION OF MARS
SEARCHING FOR THE COSMIC ORIGINS OF LIFE

THE EXPLORATION OF MARS
SEARCHING FOR THE COSMIC ORIGINS OF LIFE

PIERS BIZONY

AURUM PRESS

This edition first published in Great Britain
1998 by Aurum Press Ltd
25 Bedford Avenue
London WC1B 3AT

Copyright © Piers Bizony 1997, 1998

All rights reserved. No part of this book may be reproduced or utilized in any form or by any means, electronic or mechanical, including photocopying, recording or by any information storage and retrieval system, without permission in writing from Aurum Press Ltd.

A catalogue record for this book is available from the British Library.

Jacket design by Clive Dorman

ISBN 1 85410 584 1

1 3 5 7 9 10 8 6 4 2
1998 2000 2002 2001 1999

Printed in Great Britain by MPG Books, Bodmin

DURHAM CO ARTS LIBRARIES & MUSEUMS	
5642410	
	11.11.98
629.455	£12.95

This book is dedicated to the memory of

Carl Sagan

CONTENTS

ACKNOWLEDGEMENTS	10
FOREWORD	11
ONE: FIRST CONTACT	**16**
Dramatic Mars	20
A Dry and Frozen World?	22
The Rivers of Mars	24
TWO: THE VIKING INVASION	**27**
Touchdown	34
The Hard Problem	37
The 'Biology Box'	44
The Gas Exchange Experiment	46
The Labelled Release Experiment	51
The Pyrolytic Release Experiment	54
The Gas Chromatograph/Mass Spectrometer	59
THREE: NORTHERN EXPOSURE	**63**
A Simulation of Life?	68
Meanwhile ...	69
A Team Divided	70
Ultraviolet Destroys and Creates	74
FOUR: THE DAWN OF LIFE	**76**
The Earliest Traces of Life	79
Peroxides Again	84
When Worlds Diverge	85
The 'Cambrian Explosion'	86
Turning Molecules into Life	90
Mars Aside	96
FIVE: METEORITES FROM MARS	**99**
Carbonaceous Chondrites	102
Organic 'PAH' Compounds	104
Shergotty-Nakhla-Chassigny Meteorites	107

Meteorite ALH84001   109
The Story Breaks   111
The Meteorite's Odyssey   112
Inside the Rock   114
Martian PAHs?   117
The Isotope Ratios   120
An Analytical Artefact   121
Yes or No? Is It or Isn't it?   123
Losing Ground   124

SIX: RETURN TO MARS   127
Mars Pathfinder   131
Sojourner   134
What Next?   138
A Russian Catastrophe   142

SEVEN: LIFE FROM DEEP SPACE   144
The 'Tunguska Event'   146
Life from Comets?   147
The Methane Mystery   149
Life in Space   150
Why Space? Why Not Here?   153
The Competition Problem   154
Replicating the Design Data   159
Partial Panspermia   159

EIGHT: DRAKE'S NUMBERS   164
The Greenbank Equation   165
Other Suns, Other Worlds   168
Where Are They?   170
The Likelihood of Solitude   170
Galactic Central   174
The Drawbacks of Contact   175
The Limits of Intelligence   179
The End of Evolution   181

NINE: AN ORDERED UNIVERSE?   183
The Anthropic Principle   186
The Inevitable Universe   188

TEN: THE LIMITS OF SCIENCE   191
Why We Cannot Know the Truth of Things   193
Life Into Dust and Dust Into Life   195

FURTHER READING   197

Reproduced by kind permission,
KAL/Writers & Cartoonists Syndicate

I WOULD LIKE to express my gratitude to a number of people who assisted me with this project. They found the time to answer my queries, or send me material. In alphabetical order, they are: the Armagh Planetarium, Dr Everett K. Gibson, at the Johnson Space Center, Houston, Texas; Dr Monica Grady, at the Department of Minerology, The Natural History Museum, London; Mat Irvine, of the Duster Production Company; Dr Roger Launius, at the Nasa History Office, Washington DC; Dr Gilbert V. Levin, at Biospherics Inc., Beltsville, Maryland; Dr David S. McKay, at the Johnson Space Center, Houston, Texas; Douglas Millard, Curator of the Space Gallery, at the National Museum of Science and Technology, London; Frederick I. Ordway, at the Space & Rocket Center in Hunstville, Alabama; Prof. Colin Pillinger, at the Department of Planetary Sciences, The Open University, Milton Keynes; and Dr David G. Stork, at the Ricoh Research Center, Menlo Park, California.

Piers Bizony

FOREWORD

WHEN THE ITALIAN astronomer Giovanni Schiaparelli drew the first map of Mars in 1877, he had nothing more to guide him than the blurred images captured by his simple telescope. He recorded, as best he could, several large dark plains loosely connected by much narrower features, which he labelled 'canali'. In Italian, this simply means a groove or channel. An enduring scientific myth was created when Schiaparelli's work was translated for English-reading astronomers. 'Canali' was interpreted to mean 'canals', artificial water-bearing structures created by an intelligent civilization.

In 1894 a brilliant but misguided American astronomer, Percival Lowell, made observations of Mars from a high-altitude observatory in Flagstaff, Arizona. He was sure he could see Schiaparelli's 'canali' criss-crossing the entire Martian surface. In 1896 he published his findings: 'On Mars we see the products of an intelligence. There is a network of irrigation .... Certainly we see hints of beings in advance of us.'

What might Lowell have 'seen' if he'd never encountered those flawed translations of Schiaparelli's findings? We'll never know. Lowell's work was derided by other astronomers, but his misapprehensions turned out to be much more entertaining than the reality. His concept of an ancient water-starved civilization girding an entire world with canals and pumping stations was too good to waste. Newspapers and popular writers took up Lowell's ideas with gusto—not least H.G. Wells in his classic novel *The War of the Worlds* (1896), the famous tale of an invasion of Earth undertaken by intelligent but merciless Martians. The novel's opening lines retain all their power today:

No one would have believed in the last years of the nineteenth century that this world was being watched keenly and closely by intelligences greater than man's and yet as mortal as his own; that as men busied themselves about their various concerns, they were scrutinized and studied, perhaps almost as narrowly as a man with a microscope might scrutinize the transient creatures that swarm and multiply in a drop of water .... Across the gulf of space, minds that are to our minds as ours are to the beasts that perish, intellects vast and cool and unsympathetic, regarded this earth with envious eyes, and slowly and surely drew their plans against us.

By the end of the 19th century, astronomers had acquired a new tool, the spectroscope. Analysis of sunlight reflected from the Martian surface, or glancing obliquely through its atmosphere, provided some sobering data. The air was whisper-thin, incredibly cold, and comprised mainly of carbon dioxide. Neither Lowell nor the fiction writers allowed these observations to interfere with their grand visions of an intelligent civilization. The dominant 20th-century image of Mars was promoted in cheap story magazines. Edgar Rice Burroughs, creator of the jungle hero Tarzan, wrote a series of fantasies set on Mars (called 'Barsoom' by its inhabitants). Just as Wells had been, Burroughs was greatly influenced by Lowell's ideas. Through his cheap and cheerful stories, a wide reading public soon developed a taste for life on the Red Planet. Barsoom's canals were the least of its attractions. It also boasted a substantial supply of young princesses.

Mars hit the headlines in 1938 when Howard Koch and Orson Welles adapted H.G. Wells's *War of the Worlds* as a radio

FOREWORD

play, which included a pseudo-realistic news broadcast. Many Americans thought the bulletin was genuine, and that Martian invaders had landed in New Jersey.

During the 1950s most scientists reconciled themselves to the fact that Mars was not a likely abode for intelligent forms of life. However, they still expected *something*. As the space planners in America and Russia developed their new and potent rocket technologies, Mars became a favoured target. A manned exploration seemed possible, in theory at least, for the next generation. In *Das Marsprojekt* (1952), the German rocket engineer Wernher von Braun outlined how 1950s technology might achieve a mission. Ten ships, 4,000 tons apiece, would make the trip. His ideas were incorporated into a grand vision of space exploration, popularized in family magazines and books. These helped to establish the political groundwork for the founding of the giant American space agency Nasa in 1958.

Arthur Clarke wrote a fine science fiction novel, *The Sands of Mars* (1951), depicting a self-sufficient colony of earthlings breaking away from Mother Earth—a theme that continues to haunt scientific and literary imaginations. Ray Bradbury used the planet as the setting for one of the most famous science fiction works of the 20th century, a poignant novel, *The Martian Chronicles* (1950), which was rather more about us than 'them'. His Mars was inhabited by the ghostly remnants of an alien culture. Bradbury also depicted the carelessness of humans when they enter a new and unknown environment. Robert Heinlein promoted his hopes for the betterment of mankind by using Martian intelligence as an exemplar in a major novel, *Stranger in a Strange Land* (1961). Its influence fed into the 1960s counter-culture.

As space flight became ever more feasible during the 1950s, the Hollywood film and television industry responded with some Mars-related films of varying quality. George Pal made a superb version of *War of the Worlds*, along with a couple of less dramatic but technically impressive space exploration films, including *The Conquest of Space*, which centered around a von Braun-style Mars mission. Walt Disney produced an influential series of television documentaries, *Man in Space*, under the keen guidance of von Braun. The third segment detailed the exploration of Mars using a fleet of support ships and giant gliders for landing.

So when did the current version of Mars first take hold of our imaginations: the red and lifeless, boulder-strewn world we *think* we know today? By the 1960s, Lowell's canals were firmly ruled out. However, in a report called *Conquering the Sun's Empire*, prepared by the American space agency Nasa in 1963, the authors stated, 'We can feel reasonably confident that primitive life exists on Mars.'

The first tiny and fragile space vehicles flew near the planet in the mid-1960s. A decade later, complex robot landers were sampling the Martian surface soils, using instruments designed specifically to detect life.

In July 1996, a team of Nasa scientists announced that they had discovered the possible traces of fossilized microorganisms in a meteorite originating from Mars. They are eager to prove the existence, on another world, of 'transient creatures that swarm and multiply in a drop of water'.

This book tells the story of Martian exploration: the space vehicles, the experiments, the recent analysis of Martian meteorites that fell to earth, and the intense debates surrounding all these scientific activities. It seems we are just as fascinated with Mars *now* as we were a century ago.

However, every question about the possibility of microbes on Mars leads to greater questions about life as a whole. I've tried to write a compact synthesis of the broader debate about life on earth, on Mars, in the solar system, and in the galaxy at large. Where and how did life begin? Was life on earth 'seeded' from space? Are there any intelligent alien civilizations within neighbouring star systems? If so, do we have a reasonable chance of communicating with them? Will they visit us? Will we visit *them* one day?

The final chapters deal with the cosmos as a whole, and its possible significance. Is our physical universe a chance creation of random forces, or is there something about the laws of nature which stacks the odds deliberately in favour of the emergence of life and intelligence?

FOREWORD

The quest for our cosmic origins takes us on a fascinating journey through all realms of time and space, from the subatomic to the infinite. It encompasses chemistry, biology, evolution, philosophy and morality. We start with a supposedly simple scientific question—does Mars harbour primitive microscopic organisms? This leads us, step by step, to the deepest question of all. What is the meaning of *our* lives?

CHAPTER ONE
FIRST CONTACT

RUSSIAN SPACE PLANNERS could be forgiven for believing that Mars has a long-standing grudge against them.

Despite beating America into space—launching the first satellite and orbiting the first man—flying a simple robot probe to one of our nearest neighbours in the solar system has proved a more elusive challenge. Not that the Russians haven't tried. Between October 1960 and July 1987, they attempted 16 unmanned missions to Mars. However, scarcely a single flight was completed without mishap. Six of the designated booster rockets fell at the first hurdle, never even getting their payloads out of the earth's gravitational field. Two barely left the ground.

Nevertheless, the Russians earned the distinction of being the first to send an artificial object into the broad vicinity of the sun's fourth planet. In June 1963, Mars I (strictly speaking, the fourth probe after three failures) approached to within 193,000km of its target.

Even so, the probe's mission controllers were in no mood to celebrate. All radio contact had ceased three months earlier, and the mission was a total wipeout. Another probe skittered past Mars in February 1974, but missed its chance to brake into orbit. All it could manage were a few hasty over-the-shoulder televi-

FIRST CONTACT

sion pictures before disappearing for ever into the depths of space, far out of range of its target.

Three probes carried additional lander capsules. All failed. Mars II crashed in November 1971. A month later, Mars III unwisely tried to descend to the surface during a planet-wide dust storm. Its vertical rate of descent was probably gentle enough, but its horizontal velocity must have been breakneck as the Martian wind scooped up its enormous parachute. Russian astronomers on earth watched through powerful telescopes as the storm built up, but there was nothing they could do to avert a crash. The electronic 'event sequencers' aboard Mars III had been programmed months in advance. Everything was fully automated. The descent could not be halted. The instrument capsule may well have been dragged for hundreds of metres across the rocky terrain before coming to rest, tilted at a crazy angle and all but smashed to pieces. This would explain why it transmitted just 20 seconds of television signals—they could hardly be described as pictures—before falling silent. In March 1974, the Mars VII lander somehow ended up missing the ground altogether by more than 1300km.

America's Nasa space agency has had better luck. Of their nine Mars probes launched to date, only three have failed.

On July 15, 1965, Mariner IV flew past Mars after a trouble-free journey of 230 days, coming to within 10,000km of the planet at its closest approach. In these earliest days of remote imaging technology, the transmission of pictures from deep space was no simple matter. Mariner's camera recorded 22 images onto a strip of photographic negative film, which was then processed internally on a convoluted series of rollers. A scanning device, similar in principle to a fax machine, then translated the results into radio signals for the waiting earth, with pulses corresponding to light or dark areas on the negatives. Each picture took eight hours to transmit. In modern computer terms, this would represent a snail-like rate of eight bits per second. In order to send the entire batch of 22 images, Mariner's radio gear beamed away for ten days solid. The pictures' black-white values were then electronically 'reversed' by the earth-based receiving equipment to produce positive images: the first close-up look at Mars, in fuzzy frames of 200 scan lines, each consisting of a string of 200 tiny dots.

The pictures were a great achievement, but somewhat disheartening all the same. Of course, nobody had seriously expected Mariner to encounter Percival Lowell's famous 'canals', let alone the romantic ruins of ancient cities; but neither had the mission scientists expected to find craters—hundreds of them. Judging from these first views, it seemed that Mars might be just like the moon, pockmarked all over with impact scars, and just as inactive. At the very least, there had been some hope of finding gentle undulating plains, dune-swept deserts and signs of dynamic erosion processes.

The craters were a distinct disappointment. Their abundance, and the relative sharpness of their definition, appeared to indicate that nothing very much had happened on Mars since the craters had been formed many millions of years ago. Even more frustrating, there were no obvious traces of vegetation. Patterns of dark areas in the equatorial regions of Mars had often been observed from earth through telescopes. They were just as often attributed to vast tundras of lichen-like plant forms clinging to the beds of dried-out seas. Now, they were revealed as nothing more than the most subtle differences in the shading of geological features.

Since the 1870s, successive generations of astronomers had recorded apparently gradual changes in the shape of some of these dark areas, all of which had been attributed to seasonal spurts and fade-outs of plant growth. Mariner IV found nothing to support these intriguing ideas. Dust storms, rather than life forms, were responsible for the illusions of widescale surface change.

However, Mariner IV's reconnaissance had covered only a tiny proportion of the planet's surface. Proponents of Martian life were not yet willing to abandon all their hopes. Such a small selection of fuzzy pictures could hardly be considered definitive. Who could tell what a broader mapping survey might reveal?

During 1969, two more Mariner-class probes flew past Mars, returning 170 images between them. After the drab disappointments of the first fly-by, the new pictures revealed that there was, happily, much more to the Martian surface than just craters. There were also large areas of smooth terrain, especially in the northern hemisphere, which the earlier probe had largely failed to photograph. Mariners VI and VII also confirmed beyond doubt the existence of huge ice caps at the poles, made up of frozen carbon dioxide and water ice. In addition, a few pictures showed what appeared to be wispy clouds in the atmosphere, and early morning fogs nestling in the lee of crater walls. Clearly, there was a weather system here. Mars was revealing itself as a dynamic planet, with its own equivalent of earth's summers and winters. Those dust storms may have disappointed the scientists hunting for lichen, but nothing could alter the fact that they were *seasonal*.

These fleeting fly-bys were as much frustrating as they were fruitful. They were also wasteful, with expensive spacecraft beaming back a few days' worth of data before their hurtling trajectories carried them too far away from Mars to be of any use. Nasa devised a more advanced variant of their Mariners, with small rocket motors that would brake them into a stable orbit so that they could stay on hand for long enough to survey the planet in detail and across a much wider area. It was hoped that a more or less complete map of Mars could thus be obtained. The television camera system was redesigned, with lenses of different focal lengths for wide and close shots, a three-fold increase in image resolution (700 scan lines), a double-jointed motorized platform that could point the cameras in many different directions at the behest of ground controllers on earth, and a suitably more robust system for storage and transmission of the resulting data.

Two of the improved Mariners were launched, three weeks apart, during May 1971. Nasa intended a complex double flight, with one probe orbiting over the Martian poles, so that the rotation of the planet below would eventually bring its entire surface under the mapping cameras. Meanwhile, the second probe would adopt an equatorial orbit, passing repeatedly over the same belt of territory. In this way, any subtle changes in surface temperatures and features occurring over time could be closely monitored.

THE EXPLORATION OF MARS

Unfortunately, the first of the twins was lost during launch, finishing up at the bottom of the sea 500km north-east of Puerto Rico. Nasa aimed the survivor, Mariner IX, to hit an oblique 'compromise' orbit. It reached Mars after 167 days, attaining orbit on November 14, 1971. After its journey of 390,000,000km the probe ended up off-target by a distance amounting to rather less than the length of Manhattan Island. Nasa had achieved a navigational accuracy across that colossal span of space of better than 0.00001 per cent. The spacecraft systems were all working. The antenna was beaming back data. The cameras were on ...

... but there was nothing to see.

Mariner IX arrived neatly on time for the same dust storm that crippled Russia's Mars III landing effort. While the Russians could do nothing to prevent their vehicle going through its programmed motions, American mission controllers now enjoyed the benefits of what they called an 'adaptive' technology. The on-board event sequencers could be reprogrammed by radio signal. Mariner IX was instructed to wait in orbit until the storm had subsided. The cameras scaled down their rate of firing so as not to waste resources. It may not seem much today, but there was a significant advance implicit in this new ability to command changes of behaviour, as events unfolded, in a robot spacecraft 90 million kilometres away from earth.

In the event, Mariner IX had a very long time to wait. The planet-wide dust storm persisted for some two months into the New Year of 1972, during which time there was nothing much more interesting to photograph than four small, dark, elliptical smudges, arranged in a 'T' shape, poking through the murk. Since the smudges did not move in relation to the planet as a whole, it seemed reasonable to assume that they were the uppermost tips of high geological features, perhaps ridges or mountains. Indeed they were.

DRAMATIC MARS

When the dust cleared at last, the smudges identified themselves as the open-ended vents or 'calderas' of four colossal volcanoes. The greatest of them, suitably christened Olympus Mons, covers an area equivalent to Arizona, is capped by a caldera that could

FIRST CONTACT

swallow the entire island of Hawaii, and climbs 27km into the Martian sky. Mount Everest, our tallest peak, reaches less than half that height.

The other dominant features discovered by Mariner IX don't climb, they plunge. To insult the reputation of Arizona once again: tourists flock to see its most impressive geologic scar, the Grand Canyon, a gorge 350km long, 24km wide in places, and something over 1.5km deep. On Mars this would count as the merest scratch. Valles Marineris, a tangled network of canyons, wraps its way around the planet for thousands of kilometres. The principal trenches are more than 200km wide in places, and their steep walls extend 7km below the crust of the surrounding plains.

The main problem with all these features, at least from the tourist's point of view, is that they are so large they can properly be appreciated only from space. The slopes of Olympus Mons actually have a very gentle incline when viewed from ground level; and the walls of Marineris are so far apart that the opposing faces often cannot be seen at the same time.

The most colossal Martian feature is also the least obvious. The major volcanoes, the Marineris trenches, and large areas of rifts and fractures are associated with the Tharsis Bulge, a vast region where the planet's sphere was pushed out of shape by the internal pressures of magma. This molten rock beneath the crust pushed Tharsis outwards like a blister. The great quartet of volcanoes was fed by the blister's hot interior. The magma's journey towards these volcanoes may have formed an underground channel whose crustal roof eventually fell in to create Marineris. Another possibility is 'tectonism': the gradual sliding apart over time of great slabs of the planetary crust. The snag here, though, is that Mars doesn't appear to exhibit substantial tectonic shifts.

The sheer size of the volcanoes associated with the Tharsis bulge provides an interesting clue. On earth, continental plate tectonism frequently moves active volcanic peaks away from underlying magma 'hotspots' and the volcanoes eventually close up. Meanwhile the hotspots push up new volcanoes as fresh crust slides slowly into place overhead. The Hawaiian islands constitute a long chain of volcanoes formed in this way. By stark contrast, Olympus Mons and its companions are huge, and quite isolated, so that they must have remained on station above their magma hotspots throughout their periods of activity. The calderas show a pattern of craters within craters, betraying a long history of multiple eruptions from the same site.

Scientists are still arguing about the sequence of volcanic events, but it is broadly agreed that the Martian interior is not quite so volatile today as it once was. The biggest indicator of this quiescence is the weakness of Mars' magnetic field in comparison to earth's. As our world rotates on its axis, so the solid crust drags the viscous, iron-rich, molten interior around with it. The net result is that the interior's speed of rotation is slightly different from that of the crust. A slow but colossal dynamo effect produces a powerful magnetic field. By contrast, Mariner found no obvious magnetic field around Mars, and from this unexpected negative result, geologists concluded that there is little internal drag. The interior may still be quite hot, but the magma is rather more sluggish today than it was between three and four billion years ago, when it pushed up the Tharsis Bulge and spewed out in seemingly endless rivers of fire from the great volcanoes.

Mars has not been shaped by its internal energies alone. The planet's geology has been affected throughout by meteorite impacts. The southern hemisphere is noticeably more cratered than the northern, but it seems unlikely that this reflects the actual impact history. One theory suggests the entire northern hemisphere was churned up by a single cataclysmic encounter with a large asteroid. Vast areas of the outer crust melted. Magma was released from beneath, and the previously crater-scarred face of the northern hemisphere was all but erased.

A DRY AND FROZEN WORLD?

Mariner's procedure for making atmospheric measurements is worth detailing. During the approach phase of the journey, spectroscopes slung underneath the vehicle looked at Mars full-face and observed the spectrum of sunlight reflected from the surface. Scientists could thus determine some characteristics of the ground by noting which wavelengths had been absorbed or scattered by Martian soils. Obviously, the reflected light had to pass through the atmosphere before reaching the instruments. To what extent had the atmosphere contributed its own absorptions and scatterings?

A second type of measurement was taken when Mariner's orbit carried it past the terminator and around the planet's shadowed side. At that juncture, all the reflected surface glare faded into night. The spectroscopes were now pointed a fraction beyond the curving horizon, with a tight field of view that only narrowly avoided the planet itself. They measured the light from the sun as it filtered through the thin sliver of atmosphere in the last moments before eclipse.

Now the mission scientists could ask themselves two simple questions: what did Mars' atmosphere do to sunlight passing straight through it? And how did the spectrum derived from that measurement *differ* from the spectrum of sunlight bouncing off the Martian surface, as obtained earlier? By comparing the data, and by using the known spectrum of pure sunlight as an absolute yardstick, Mars could be made to reveal some of its chemical constituents, both on the ground and in the sky.

Of course, spectrographic analysis has been the principal tool of earth-based astronomy for more than a century. As far back as 1867, astronomers using simple spectroscopes had been able to determine the woeful lack of oxygen and water vapour in the Martian sky, even if their measurements lacked precision. In 1926 more advanced readings provided astronomers with some idea of Martian temperatures—perhaps zero degrees Celsius on a summer's mid-afternoon, falling to minus 65 degrees Celsius at night. By 1947 astronomers had a reasonably good understanding of the atmosphere's principal chemical components. Mariner was far from unique in its capabilities, but it had one prime advantage: its measurements were taken right up close to Mars. Their sensitivity was unprecedented. Better still, the on-board spectroscopes were not confined to working with visible light. They worked within several regions of the electromagnetic spectrum, including infrared, ultraviolet, and certain radio frequencies.

THE EXPLORATION OF MARS

The Mariner program confirmed what most astronomers already suspected: that the atmosphere of Mars consisted mainly of carbon dioxide. But the density was even lower than anticipated, at barely one per cent that of earth. Any faint possibility of liquid water flowing on the planet was ruled out at a stroke. In such a thin atmosphere, prevailing gas pressures are so slight that water molecules become detached from the liquid state at the slightest application of heat energy from the distant sun. Frozen water, however, has already lost its heat. Its molecules are not in violent kinetic motion, and so will not separate. Thus, on or around the Martian surface, water can exist only as ice or vapour, never as liquid. When sunlight hits a patch of ice or frost on the ground, traces of vapour escape without first becoming liquid. The water clouds in the sky consist of tiny ice particles, rapidly condensing and freezing in the atmosphere.

In the absence of liquid water, any chances for the presence of life seemed much diminished. Scientists began to speculate about water trapped underground. Mars' core may have lost much of its heat, but maybe temperatures and pressures below the surface were still high enough to allow for flowing water. Unfortunately, Mariner was not equipped to investigate such possibilities. Nor was the resolution of its cameras detailed enough to rule out altogether the possibility of small, hardy organisms somehow surviving on the surface—not, for sure, in great swathes visible through telescopes on earth, but perhaps hiding in isolated microscopic colonies within the fissures of rocks. This idea was total speculation, the parting shot of disappointed biologists whose planet-spanning lichen tundras had vanished overnight back in July 1965.

Today, more than thirty years later, theories about bugs in boulders have returned in strength.

THE RIVERS OF MARS

Yes, it was a great pity that Mars had no liquid water. Which was odd because there seemed to be river beds. On July 1, 1972, eight months into its mission, Mariner IX transmitted pictures of Mangala Vallis, a 600km channel in the southern, cratered hemisphere. Unlike Vallis Marineris, the jagged trench which cuts through a massive plain like a geological fault line, Mangala starts off as a fine network of tributaries covering hundreds of square kilometres of an upland area. These then merge into a deep but narrow main channel that 'flows' down to lower-lying territory. As the channel reaches lower, and stretches further from the tributaries, so it seems to thicken out. Some kind of fluid must have travelled down from the higher ground, gathering mass and pace, consolidating into a single main flow, and gouging out an ever-expanding trench as it hit the lowlands. Leaving aside the fact that it has no water today, Mangala is comparable in scale and characteristics to the earth's Amazon river. Intrigued, the scientists christened the downstream plain Amazonis Planitia in honour of this uncanny similarity.

As the Mariner IX survey progressed, many such channels on Mars were revealed, with three broad variations. 'Run-off' channels are 'V'-shaped in cross-section, which indicates a vigorous flow cutting deeper and deeper into the surface over time. 'Outflow' channels seem to spring fully-formed from areas of chaotic terrain, and the best theory to date suggests that they may have resulted from sudden collapses of crust and the release of subsurface waters in great floods. The third, 'fretted' class of channel has flatter floors. Many of these channels show signs of layered sediments, which suggest the possibility of water flowing slowly enough to deposit soil particles carried along from the upland sources.

Mariner IX came to the end of its useful life after almost a year, in October 1972. The internal guidance systems which had kept the machine on station suddenly drifted out of line, and the ship lost its bearings. It never regained a stable attitude in relation to the Martian horizon, and the cameras and scientific instruments were left pointing randomly into space. The fragile communications link was compromised. Earth was a tiny target for Mariner's radio dish, a pinprick of light in an infinite and tumbling horizon.

Nasa called it a day, but nobody was complaining. A tremendous amount of data had been obtained, despite the long dust storm which had greeted Mariner on its arrival. The probe had discovered seasonal weather systems, wispy clouds and fogs, giant volcanoes, polar ice caps, a dramatic geology, with immense chasms and fractures, and maybe colossal dried-up river beds. Mars was, after all, a fascinating place.

Space scientists allowed themselves to think the unthinkable: at some stage in its 4.5-billion-year history, Mars might have supported colossal quantities of liquid water. If there had been flowing water, then by definition Mars must once have been a much warmer world than it is today, with a thicker atmosphere, otherwise the water would have boiled or frozen long before it could scar those great channels across the surface.

Did those gigantic volcanoes pump a rich and thick broth of gaseous materials into the Martian sky at some time in the far distant past? Did ancient Mars once support the rivers of life?

CHAPTER TWO
THE VIKING INVASION

IT COST $42 MILLION; it was brimful of powerful rocket engines and aerospace hardware; it gleamed white and brilliant in the summer sunshine; yet it was destined to remain firmly rooted to the ground.

On the morning of July 1, 1976, the Smithsonian Institution in Washington D.C. opened the doors of its brand-new gallery, the Air and Space Museum. A prettily coloured inaugural ribbon was stretched across the main entrance. President Gerald Ford stepped up in front of it, but he didn't make the crucial cut. A radio pulse from a tiny transmitter more than a hundred million kilometres away triggered an electric guillotine, and the ribbon was sliced by the very remotest of remote control.

Twelve days earlier, on June 20, Nasa's latest space probe, Viking I, had achieved Martian orbit after a flawless ten-month flight. The gimmick radio signal to Washington whetted public and political appetite for further treats to come. On July 4, Independence Day (in 1976 it also marked the 200th anniversary of America's revolt against British rule) the Viking mother craft was supposed to send a lander right down to the surface of Mars, thus nailing national pride yet more firmly to the pages of history. Nasa's managers were looking forward to the Viking's public relations benefits as well as the purely intellectual rewards. The moon landing missions were over, and for the first time in a decade and a half the space agency had launched no manned rockets in an entire year. Nor would there be any such missions for several years to come. The semi-reusable winged space shuttle was beginning its long and tortuous development. The famous Mission Control complex in Houston was engaged in make-work simulations for missions they couldn't yet fly. Hundreds of veteran space workers had been laid off.

At the Florida launch centre the gargantuan service towers for the Saturn moon rockets were being dismantled. Brilliant sparks still flew out from the towers at intervals, but not from rocket ignitions. They came from the scrap merchants' acetylene torches as they cut through the steel girders.

The remaining space kudos now belonged to a relatively small and little-known control centre at the Jet Propulsion Laboratory, tucked away in a suburb of Pasadena, California. Operating a robot probe might have lacked the glamour of sending men to the moon, but in many ways Viking was among the most challenging of all Nasa's projects.

Building Viking had been an intensive operation. The machine was breathtakingly complicated. Admittedly, the top half of the ensemble was familiar enough: yet another up-grade of the Mariner design, with its squat, louvered hexagonal box sprouting four solar panels. This mothership (the 'orbiter') may have been constructed around slightly old technology, but the vehicle was fitted with hefty reserves of fuel and power. Viking's lower section, however, was something quite new. It was a very smooth, white, saucer-shaped canister, split into two distinct halves like a pair of inverted soupbowls joined together rim to rim. This 'aeroshell' was coated in heat-shielding material designed to withstand a fiery entry into the Martian atmosphere, at the end of Viking's ten-month outward voyage. Yet another layer of cladding, called the 'bioshield', protected the entire aeroshell and its carefully sterilized contents against contamination. A space worker with a minor throat infection could have jeopardized the credibility of the entire project. Nasa scientists did not wish to send earthly bacteria to Mars by mistake, in

THE VIKING INVASION

THE VIKING ORBITER AND AEROSHELL

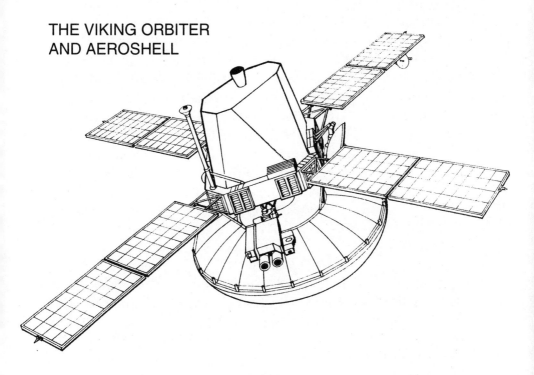

case the bugs came to like it there and took up home, destroying forever any possibility of determining beyond doubt whether or not Mars supported life of its own—Viking's primary purpose. Once the vehicle was in deep space, safely clear of earth, the bioshield would be discarded.

Inside the double-skinned aeroshell was the lander itself. In contrast to the mothership's reflective gold foil cladding and shiny solar arrays, the lander's dominant colouring was a flat, neutral grey. Secured for flight, it was like an insect larva in a cocoon, with its antennae folded into tight bundles and its three legs tucked tightly under its body. Such a comparison with a living creature is not entirely fanciful. This was a very 'smart' device, with a rudimentary electronic intelligence. The late Carl Sagan of Cornell University worked at the heart of the Viking program. In his book *Cosmos* (1980) he recalled that by some standards the lander was as clever as an insect, by others only as intelligent as a microbe. 'It takes millions of years to evolve a bacterium, and billions to make a grasshopper. With only a little experience, we were becoming fairly skilful at it.' This degree of intelligence was expensive to achieve, but indispensable if a Russian-style crash was to be avoided.

At the planned time of landing, radio commands would take at least 19 minutes to reach Mars. Allowing for the equivalent response time from earth, a 38-minute radio time lag had to be factored in. For a well-understood orbital rocket burn scheduled many months in advance (or, for that matter, a ribbon-slicing ceremony in Washington) this delay could be circumvented by transmitting the relevant commands 19 minutes ahead of time. However the Viking lander, hovering a few hundred metres above a rock-strewn surface of unknown characteristics with its last dregs of retro-rocket fuel burning away, would have a few seconds, rather than minutes, to find a safe spot for touchdown. At the Jet Propulsion Laboratory (JPL) in Pasadena, a team led by Neil Herman designed an electronic brain for Viking. This was not just another electrical 'sequencer' but a fully-fledged digital computer, coupled to sensitive radar altimeters, and quite capable of making all the necessary decisions with very little help from earth.

By the standards of early 1970s technology, JPL's self-sufficient hardware was a fantastic achievement.

The mothership, for its part, turned out to be capable of everything asked of it. Viking's outward-bound flight had gone very smoothly. With the craft safe in orbit around Mars, all JPL had to do now was to prime the lander and let it off the leash. A July 4 touchdown would have been just perfect, but there was a snag. The mission managers couldn't agree on a safe place to touch down. The prime target site was in the Chryse Planitia (the Plain of Gold), and in particular, a region where the merging of four sinuous channels cutting through the plain seemed to suggest that water had once flowed there. The old Mariner IX photos revealed this part of Chryse as a fairly flat area, relatively free of hazards. But Mariner's best photos resolved surface features down to 100 metres only. Viking's more advanced video cameras now revealed that the touchdown site was much rougher than anybody had expected. There were countless large boulders scattered everywhere, and endless cracks and

THE VIKING INVASION

gulleys which hadn't stood out so alarmingly in the Mariner surveys. The flow of water which (it was presumed) had originally carved the four main channels showed signs of an incredible violence. The land for many kilometres around was 'chaotic'. Great slabs of crust had collapsed, and there was a bewildering array of rifts, pits and knobbly protrusions which the planetary geologists in Pasadena were at a loss to explain.

In all probability it wasn't merely a quartet of tame river tributaries that had first carved out this landscape, but a cataclysmic release of underground water surging upwards through the churning crust. Perhaps a thermal phenomenon within the Martian interior had melted subsurface ice? Or maybe an ancient aquifer had burst its restraints? Whatever the explanation, the mission controllers pored over the incoming pictures, searching for a less rugged place to put down their expensive robot. They could argue the geology at some later date.

The choice of landing areas was limited to a broad, fat belt of options around the Martian equator. These were sites that passed beneath the mothership's orbit at some time or another. All circumstances being equal, where was the *best* place to land?

This had been the subject of many months of argument. Some scientists wanted to explore the rugged, cratered southern hemisphere, where the preserved gougings of old impacts could yield some data about what went on beneath the immediate soil surface. Other theories centred around the base of the smaller volcanoes, where the dirty stains of old eruptions might explain how the atmosphere was originally created from volcanic 'outgassing'. However these sites were largely of geological interest, and Viking's prime mission was the search for life. The more temperate equatorial regions were the warmest and therefore the best-suited for life, or so it was supposed.

Given the emphasis on a mid-latitude site, it seemed sensible to set Viking down onto one of the many 'river beds' where life, ancient or recent, might have left some faint trace in the dried-up soils. The 7-kilometre-deep Marineris trenches would also have been worth a shot. This was just the sort of place where water lakes might once have laid down deposits of silt. Furthermore, the atmosphere at the bottom of the trench is slightly thicker and warmer than at any other place on Mars, even to this day. Those few extra kilometres of depth have a significant effect on the gaseous pressure, and the whole region is broadly equatorial. Here, temperatures are the highest anyway. Surely Marineris should have been a prime target for a landing?

Unfortunately most of the deeper river beds, craters and trenches had to be ruled out, for the simple reason that Viking needed a wide margin of error during touchdown. The walls of Marineris presented too great a hazard, and nobody wanted the lander to topple down the steep slopes of a riverbank, or tumble into the maw of a crater. 'Flatlands' dominated the Nasa mission planners' every waking thought.

By definition, terrain flat enough for a safe touchdown is not particularly interesting. If an alien spaceship were to approach earth for a landing, its extraterrestrial operators might gaze down wistfully at the artificial-seeming towers sprouting upwards in profusion from the great cities. The strange patterns of light, and the chaotic babble of radio stations and mobile phones would arouse the dullest alien's interest. Alternatively, their probe might home in on the obvious potential of the Amazon rain forest, with its green canopy visible from space, and its humidity and balmy temperatures readily apparent. However, the best bet for a *safe* landing would be somewhere in the vast, arid, equatorial deserts—which, from an orbital vantage point, appear largely featureless (if one discounts the exquisite beauty of their wind-formed dunes). There's much of interest to be found in any earthly desert, but a random landing among the sands of the Gobi, for instance, would prove very unrepresentative of our planet's richness as a whole. All the same, the Gobi, the Sahara or the Serengeti deserts would almost certainly provide an alien ship with its first priority: a soft touchdown with plenty of margin for errors in the descent through an unfamiliar atmosphere.

Viking faced these kinds of constraints. The most fascinating of Martian terrains were almost always the most dangerous. As many as 400 geologists, flight dynamics experts and managers pored over the new images from Martian orbit, agonizing over what to do. Even the smoothest 'plains' looked far too lumpy all

THE VIKING INVASION

of a sudden, especially now the time had come to commit their spacecraft to a landing instead of simply talking about it at pre-flight planning meetings.

Despite a full-colour capability and improved resolution, Viking's cameras were not capable of showing up details less than 20 metres across. Even this level of resolution could be achieved only at the lowest point of the mothership's egg-shaped orbit, so the high-resolution scan opportunities were by no means unlimited. What small-scale boobytraps might be down there? What if the lander came down at an angle, half on top of a small boulder? An insignificant piece of rubble just half a metre tall could tear out the lander's belly, ripping vital electronics to shreds.

Another fear was soft dust—so soft that the lander might sink into it without trace. Large swathes of Martian territory had proved unusually opaque to earth-based radar surveys. Either these 'dark' areas were very rough, and the radar beams were being randomly scattered instead of bouncing cleanly back, or else they were soft and mushy, with granular surfaces that absorbed the pulses. It was impossible to tell the difference. The only certainty was that the radars couldn't pick up a decent return signal.

The reality of accomplishing the landing wasn't quite like those pretty artworks which Nasa had released in the preceding months. These had depicted the grey, three-legged billion-dollar robot nestling confidently among smooth undulating sands, with a couple of dozen harmless pebbles no larger than tennis balls scattered about at tasteful intervals. Carl Sagan did his time late into the night, staring endlessly at the new pictures and wondering if the Viking might be 'condemned, like the legendary Flying Dutchman, to wander the skies of Mars for ever, never to find safe haven.' His colleague, Gentry Lee, wailed to a *National Geographic* journalist, 'What if we land on a boulder? What if we sink into dust? What if we hit the side of a crater? This thing can't operate at more than a thirty-degree slope!'

Day after day the world's newspapers carried ever-shrinking reports about Viking's progress at the bottom of the front page, or maybe on an inside page. Or eventually, buried right at the back. 'Mars landing delayed again.'

Fortunately, there was one big factor protecting the project against total disaster: a whole spare spacecraft. Two ships had been launched atop a pair of unmanned Titan-Centaur rockets, three weeks apart, on August 20 and September 9, 1975. Because of its slightly different trajectory, Viking II was still just over a fortnight away from Mars on July 20, 1976, the day when Nasa at last decided to try for a touchdown with Viking I.

Senior managers privately acknowledged that the entire landing effort was a fifty-fifty deal. They'd be lucky to get one of the ships down intact and would probably lose the other. There was really no point in waiting any more. The dangers of the Martian terrain were never going to change, but exhaustion among the mission controllers in Pasadena, primed day after day for a landing and disappointed every day, greatly increased the likelihood of human error or misplaced fatalism in the descent preparations.

A final show of hands among the mission controllers settled the endless disputes about the touchdown site. Chryse Planitia still came out as the best bet. However the new site, selected on safety grounds, was far removed from the confluence of those four river beds which had first attracted everybody's interest.

At 12:47 a.m. Pasadena time, July 20 (seven years to the day since Neil Armstrong and Buzz Aldrin made the first landing on the moon in Apollo XI) flight controllers instructed Viking I's lander to separate from its mothership and begin its descent. All being well, the vehicle would be on the ground in three hours and ten minutes.

TOUCHDOWN

The lander's electronic brain had a lot of work to do. Shortly after separation from the mothership, it had to flip its smooth protective aeroshell around, point four small rockets at exactly the right angle, and fire them for 27 minutes in order to decelerate from orbital velocity. Mars' gravity would then drag the machine downwards for two hours in a leisurely but precise arc.

Once the aeroshell hit the atmosphere, its curved heatshield would have to survive frictional temperatures of up to 1500 degrees Celsius. Within eight minutes, atmospheric drag should have slowed the shell to less than 900 kilometres per hour. Nasa had never tried this particular stunt before with any of their robot probes. At 7000 metres above the surface, the heat shield was designed to split cleanly apart so that the eerie grey machine inside could sprout a huge main parachute and stretch out its three legs, like a bug emerging from its chrysalis and taking wing. Then, the retro engines were supposed to fire up and balance the lander atop three precisely synchronized columns of fire. With the ground less than 800 metres away, the computer would have about 55 seconds to get everything down in one piece. It had to switch the radars to maximum sensitivity for the crucial last few metres of descent, avoid big rocks, and dump the 'chute so it didn't fall on top of the vehicle later on.

There was no time to lose after touchdown. The mission directors had in mind the Russian experience of December 2, 1971, when the Mars III probe had fizzled out just 20 seconds after touchdown. Within 25 seconds (and still operating entirely without human guidance) Viking had to deploy its antenna dish and meteorology boom; beam its position and status to the orbiting mothership; relay an 'arrived safely, wish you were here' message back home, and immediately start up a camera. If something went wrong with the lander at this stage, at least one photo of the surface—the 'contingency scan'—might thus be retrieved before blackout, along with a few seconds'-worth of temperature and pressure readings from the weather sensors.

Because of the radio delay, the Pasadena controllers had to wait for 19 minutes after the event for Viking's first signal from the surface. No postcard from any tourist was ever so devoutly wished for. At the predicted time, dozens of people squeezed into the Pasadena control room, waiting for the lander's signal. When flight controller Richard Bender shouted, 'We have touchdown!' the whole room erupted. Gentry Lee wept with joy. He and hundreds of his colleagues had worked non-stop for eight years to achieve this one thing.

At last a proper man-to-machine dialogue could now be established with the lander. Inside an hour, the first black-and-white contingency image of the Martian surface came back, strip

THE EXPLORATION OF MARS

by strip: a close-up of a couple of square metres around one of the footpads. Balance sensors showed that the lander was tilted a few degrees off the level, and further shots of the ground revealed why. One of the three round footpads was firmly perched atop a small rock, but another was buried up to its 'ankle' in fine sand, so that the pad couldn't be seen at all. Would it sink further, up-ending the whole machine, as Gentry Lee had feared? No, the worst was over. Long minutes passed, and Viking continued to transmit confirmations of its stability. The sensors indicated no sudden lurches, no sickening jolts.

Now it was safe for Pasadena to steer Viking's twin cameras to take in a broader view, for a proper and more leisurely look at the landing site. Rick Gore, the *Geographic*'s man on the spot, sidled up to the imaging team's workstation. 'The cameras didn't work this well on earth!' exclaimed team leader Thomas Mutch. 'It's just beautiful! I ... I'm sorry. I don't feel like talking right now.'

'We were lucky', Gentry Lee said, as the wide-field picture came to completion on Pasadena's monitor screens. Nobody needed to ask him what he meant. Just 10 metres away from the lander there was a very nasty 2-metre-tall boulder. They nicknamed it 'Big Joe'.

The landscape was extraordinary, with small chunks of rock scattered around. *Everywhere.* In an area specifically selected on safety grounds for its smooth blandness, Viking's geologists reckoned they could identify 30 different rock types.

The next few hours were spent checking the Viking lander's systems in general, making sure that it was fit for a long tour of duty. The seismometer, a sensitive device intended to measure 'Marsquakes', had been protected against the shock of descent by secure locking mechanisms. They were so secure though, that they wouldn't *unlock*, which was a real nuisance. But everything else seemed fine.

By the second day the cameras were ready to try a few colour shots. At first there was some confusion about the balance of hues. Did the relative intensities of blue, red and green need to be adjusted to render the scene in its 'true' colours? No, the shots were good. A small Kodak colour chart was strategically

THE VIKING INVASION

positioned on the lander's upper body. During several of the early camera scans the chart was kept carefully in shot, albeit slightly out of focus and tucked away near the edges of the frame. When the images were downloaded into the earth's computers and adjusted so that the chart's colours came out looking 'nominal', Pasadena could be sure they had obtained a natural-looking view overall. And what a view it was.

The ground was rusty-red, just as everybody had expected. This was the Red Planet, after all. But the sky! Given the thin atmosphere, it was a hundred times brighter than anyone had expected, and suffused everywhere with a creamy orange-pink haze. One might almost have mistaken the effect for a cloudy, overcast day, except that Mars cannot support any dense cloud layers. In fact a permanent suspension of very fine dust particles was reflecting the sunlight, giving the air its gentle glow. This was interesting in itself. Throughout the lander's period of operation that glow never disappeared, at least not during the hours of daylight. Even in the absence of specific dust storms, a minimum quantity of particles obviously circulated in the atmosphere on a permanent basis.

The newspapers happily printed these amazing pictures on their front pages, but some commentators thought that Nasa must have got the colours wrong. After all, everybody knows that skies are blue, don't they?

THE HARD PROBLEM

From the very start of the Viking program eight years earlier, the landing problem had been reasonably well understood, and now they'd pulled it off. In a sense, this was the least of the mission's challenges. It was just a matter of breaking the job down into discrete components, phrased in terms of specific sensor readings. What's the height above the ground? X, as determined by radar. What's the rate of descent? Y, as calculated from shifts over time in the radar results. What's the braking thrust? Z, as determined by readings of the rocket engines' temperatures and pressures. X, Y and Z could be quantified as values in a fairly simple algebraic equation loaded into the guidance computer.

The key point is that Neil Herman and his colleagues at JPL

had known exactly how to model their task in digital form when they built the navigation systems. The biologists could only dream of having such a simple problem to solve. Back in November 1971, JPL's director Bruce Murray released a simple document stating Viking's principal purpose. Everyone already knew it of course, but seeing it in black-and-white made the whole thing seem more real: 'The primary objective of the mission will be the direct search for microbial life on Mars.'

How could anybody quantify or model *that* problem in terms appropriate for Viking's electronic systems? What exotic (and very small, lightweight) new instruments, and what kind of data delivered from them, might produce suitable X, Y and Z values for the detection of alien life? It was impossible, for instance, to guess whether or not life on Mars would share any characteristics with life on earth. But if it didn't, then Viking's tests would be quite meaningless, since there would be no sensible yardsticks against which to compare them.

So, where to begin? A hunt for animals or plants, or indeed any multi-celled organisms with a visible structure seemed quite unrealistic. Viking's camera system was not expected simply to be able to take snapshots of Martian life forms crawling about the place. In the wake of the Mariner program, no one seriously anticipated anything more complex on Mars than single-celled organisms. At this tiny scale, it was better to focus on the kind of measurable *chemical* activity that Martian microbes might demonstrate, thus betraying their presence indirectly. However, when it came to predicting exactly what tell-tale traces to look for, some pretty sweeping guesswork was required.

Just look at the problems of analysis we face closer to home. There is a wide variety of terrestrial microorganisms whose chemistries use neither plant photosynthesis, nor animal oxygen respiration. Indeed, there is a whole class of 'anaerobic' bacteria which die in the presence of free oxygen. Viruses are a further complication. Like all living things they contain nucleic acids which govern their replication. They multiply in the right conditions (as parasites within other organisms), but they don't respire, they don't eat and they don't excrete. Consequently, nobody is quite sure whether or not to regard viruses as living

THE VIKING INVASION

entities. What if Viking found something on Mars with virus-like qualities, neither truly alive nor completely inert? Could any instrument find such things if they didn't give off chemical exhaust fumes? To round off the list of problems, there was the obvious possibility that Mars might harbour forms of life totally beyond human recognition. Something quite literally alien to us.

The quest had to start somewhere, though. Viking's team of scientists focused upon the one sure-fire common denominator of all life on earth: organic carbon-based chemistry. Living things absorb, manufacture and expel organic carbon-based compounds. Even viruses are organic in their construction and replication. Viking would cut to the essentials and test for any chemical activity on Mars that involved carbon. In a sense this was a bold decision, but many biologists share an intuitive belief that organic chemistry is probably fundamental to the universe as a whole. Carl Sagan admitted in 1975 that it was hard not to be a 'carbon chauvinist'.

As the senior Viking scientist, Gerald Soffen, explained, 'Carbon is incredibly flexible. The atoms can make long chains, and they can attach to other atoms in an endless number of configurations. Only carbon can provide the incredible variety of molecules needed by any living organisms we can conceive of.'

For want of any better ideas, this bias was incorporated into the mission brief, and the long process of phrasing the right questions could begin. The biology team, led by Harold Klein at Nasa's Ames Research Center in California, considered how to address the following investigations:

1) Does anything in Martian soil exchange gases with the atmosphere?
2) Does anything in Martian soil release carbon?
3) Does anything assimilate carbon?

Of course, organisms do not take up or release carbon in a pure state. The questions really apply to carbon bound up in organic molecules, or in simple metabolic waste products, of which the most obvious example is carbon dioxide (CO_2). But as we have discussed, many terrestrial microorganisms, especially the anaerobes, excrete other substances. The first question would consider this possibility on Mars with a wider search for bacterial waste products, such as hydrogen, oxygen, nitrogen and methane (CH_4), as well as carbon dioxide.

All these questions could be investigated for soil in its native state, and also for soil exposed to nutrients supplied by Viking. It might seem odd that Nasa should send food along to Mars. After all, if there were life on Mars, surely that life would already be munching its way through a native supply? Why should Viking bring along extra provisions, which in all probability might not be to the Martian bugs' taste? Why complicate the tests by contaminating the soil with substances from earth?

There were two good reasons for hauling food to Mars. First, much might be learned about the soil by seeing how it reacted with non-Martian compounds of known origin; and secondly, the biologists felt that organisms on Mars might be wholly or partially inert because of the harsh conditions of their home world. Many microorganisms on earth can hibernate for tens, hundreds and even thousands of years when food is short, yet quickly revive when conditions become more favourable. Perhaps something similar applied on Mars? Perhaps the native organisms would awaken like sleeping princesses at the touch of Viking's chemical kiss ...?

It was worth a shot, but if hibernating Martian microbes were to be tempted out of their inertia, Viking's gift of nutrients would have to be very rich indeed. Every conceivable organic compound an alien microorganism *might* thrive on was designed into the broth's recipe of vitamins, sucrose, lactose, sugars and amino acids, all of which were dissolved in a small quantity of water. ('We could live on that stuff,' said Gerald Soffen.)

So if the Martian soil interacted with the broth in some way, would this prove the presence of life? Not necessarily. A number of inorganic compounds can produce very life-*like* reactions on contact with organics.

If anything happened when the nutrients were added to Martian soil, a clincher question would have to be addressed:

4) Does Martian soil contain its own organic compounds?

If the soil took up carbon from Viking's nutrients or the

THE VIKING LANDER

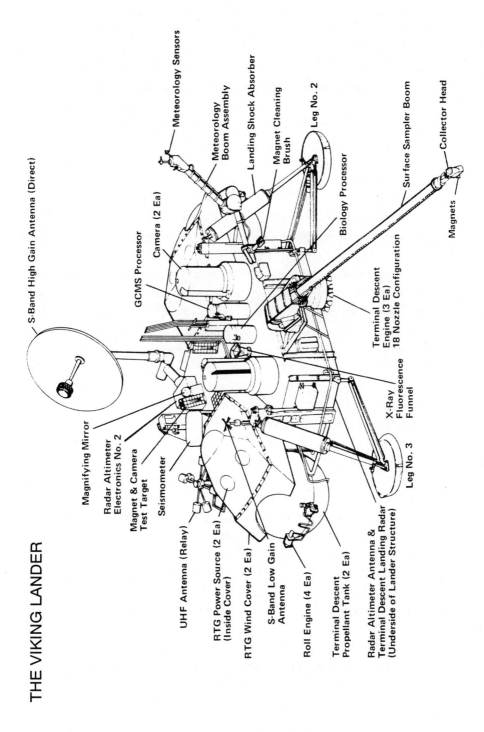

atmosphere, or delivered carbon waste products, *and* was found to contain its own organic substances, then the presence of life might be assumed. Given the unknown behaviour and predilections of Martian life, and the inevitable uncertainties of the test procedures, questions *1*, *2* and *3* could conceivably get away with one or two positive answers between them. Question *4*, however, was absolutely non-negotiable. Even if *1*, *2* and *3* were answered with a resounding 'yes', Viking's scientists would remain unwilling to conclude the presence of life if the answer to *4* was 'no'. There had to be organics in the soil, or all bets were off.

With such reasoning, the search for life was whittled down to the kind of investigation which an automated laboratory might be able to make on the surface of Mars. The hunt was on, not for qualitative or vague 'signs of life' but for specific and quantifiable chemical data about carbon chemistry: X, Y and Z. However, one of the Viking scientists, Wolf Vishniac, thought that such data, though entirely logical, was dangerously specific. He devised a fourth experiment, nicknamed the 'Wolf Trap', which made no chemical analyses whatsoever.

In the Wolf Trap, a small soil sample was dropped into a transparent jar, and a translucent nutrient fluid was added. Just one type of measurement was taken, before and after the fluid was added, and then at regular intervals over time. How translucent does the nutrient fluid remain? A calibrated beam of light passing from one side of the jar to the other was all that was required. Vishniac argued that if the fluid clouded up at any stage, then some kind of microbial growth within the nutrient could be assumed. He did not think it necessary, or even desirable, for Viking to try and investigate the exact chemical details.

Vishniac's colleagues thought that the transparent nutrient solution inside his Wolf Trap might drown its Martian guests. Any organisms still living on Mars, they reasoned, must have adapted by now to the acute shortage of water. The nutrient hoppers on Viking were thus designed to humidify, rather than soak, the Martian soil samples. In the event, budget constraints cancelled the Wolf Trap project before it could be incorporated into the Viking experiments package. Vishniac was bitterly dis-

THE VIKING INVASION

appointed, but he thought the Viking team might still find some comparative data from earth useful. In November 1973 he set out to test his trap in a place on earth that approximated, as closely as our planet can manage, to the harsh conditions on Mars: the Antarctic. Vishniac selected an intensely cold and arid valley in the Asgard Range, close to Mount Balder, in the west of the ice-bound continent. Here, he placed a selection of his traps into the ground. He had mixed a 'low-calorie' nutrient suited, he believed, to such a tough environment. On December 10, 1973, he set out to retrieve his samples. Sadly, this was the last time anybody saw Wolf Vishniac alive. At some point on his solitary walk he must have lost his footing. Later that day a search party left the McMurdo research station to look for him. His body was found at the base of an ice cliff.

Vishniac's grief-stricken colleagues determined to retrieve his samples on his behalf. The results were astounding. His nutrients were fogged with microbes, from an apparently sterile Antarctic region not hitherto considered suited to life. Vishniac's widow, Helen Simpson, identified a new species of yeast unique to the Mount Balder area. Other than scaling down the richness of the nutrient in keeping with the scarcity of food in the target site, Vishniac had made no assumptions about Antarctic biology. He had simply designed an experiment to find some. If Helen Simpson had not been on hand to identify the yeast, a suitable light beam sensor could have registered the clouding in the solution and beamed back a message to a remote base. That message would have read: 'Life is here.'

On the whole, Nasa's managers tend to design complicated projects. There's some hard-headed political reasoning in this. Simple experiments and cheap spaceships don't involve a sufficient number of researchers and manufacturers. Consequently it can be hard to obtain the backing for small-scale missions. By devising expensive and sophisticated procedures and hardware, Nasa ensures that many different organisations can profit from space activities, either scientifically or in terms of construction contracts. In this way, a moderately wide spectrum of political support for funding can be achieved. Such logic might partially explain why Vishniac's charming, simple experiment was dropped as a result of budgeting problems. The Wolf Trap was too cheap rather than too expensive. But there might also have been an image problem. It would have been difficult for Viking's political managers to justify sending a test to Mars whose result, at best, would have been: 'Yes, there's some kind of life. The nutrients went a bit cloudy.' The general scientific community would want much more rigorous data to fuel their pet theories. X, Y, and Z.

Well, they got plenty of data. Viking was equipped to undertake a wide and detailed variety of measurements, and this very breadth of analytic capability was to prove deeply confusing when the first results were transmitted to earth.

Still, the broad decisions were made. The three biology 'questions', along with all their flawed assumptions based upon terrestrial examples, were set in stone by the many financial constraints and ponderous decision-making processes within a major space program:

1) Does anything in Martian soil exchange gases with the atmosphere?
2) Does anything in the soil release carbon?
3) Does anything in the soil assimilate carbon?
4) Does the soil contain its own organic compounds?

THE 'BIOLOGY BOX'

The TRW company was contracted to build a machine that could answer these questions automatically, while operating a hundred million kilometres away from the nearest scientist. Somehow, the equivalent of a biochemistry department in a well-funded university had to be packed into a space less than the size of a car's battery, and made to work without technicians and faculty chiefs. Nearly 40,000 separate components were miniaturized to an extreme degree. By the time that Viking's principal manufacturers at the Martin Marietta company had installed the guidance computers, rocket engines, fuel tanks, miniature nuclear power plants, camera systems, temperature regulators, radio gear and antennae, there was very little room aboard the lander for anything else.

The first problem for the biology box was strictly mechanical: how to collect soil samples and get them to where they were needed. Viking's robot arm was designed to scoop up samples from around the lander's feet. It could stretch more than 2 metres outwards from the ship's body. The arm's long boom was constructed from parallel strips of stainless steel, 'kinked' for rigidity along their entire length. The narrow mouth of the storage canister flattened the kinks so that the boom could be rolled up like a glorified tape measure. It was clever, but no more so than it had to be. (In the last couple of decades, American space engineers seem to have lost much of their old clear-sightedness when it comes to designing the moving parts of space probes.)

Once the scoop was filled with soil the boom's metal ribbons retracted, shrinking the arm to its minimum length. In a clever display of double-jointedness the arm now swivelled about-face so that its scoop was sitting over the lid of the Viking lander; specifically, over a hatch on top of the biology compartment. Underneath the hatch there was a hollow cylinder, partitioned into six wedge-shaped compartments. As the scoop dumped its contents through the hatch, the cylinder rotated and each wedge received some soil. After the outer hatch was closed, the cylinder swivelled once again, dumping the contents of its wedges into yet more receptacles deep inside the laboratory.

In this way, portions from the original scoopful of soil could be distributed to different experiments. These were housed in sealable subcompartments with yet more 'carousels' inside them. The entire experiments package was reminiscent of those Russian dolls which open up to reveal more dolls-within-dolls. The point of this mechanical complexity was to ensure that all the experiments obtained samples that were as alike as possible. Another consideration was the possibility of mechanical failure in the robot arm. As long as it worked the first time at least, then all the experiments should receive some soil.

Nutrients could now be added, and their effect on the soil measured. But first, a slight complication had to be dealt with. With all the attention focused upon carbon, the nutrient's organic molecules had to be 'labelled' in some way, so that Viking could distinguish earthly carbon in the nutrients' organic compounds from its Martian equivalent. All carbon atoms have six electrons and six protons, but their isotopes come in several forms, defined by different numbers of neutrons in their atomic nuclei. (Neutrons are not involved in binding with other atoms, and they carry no electric charge.) Carbon-13 and carbon-12 are the common stable isotopes. Biochemistry makes use of both isotopes, but there is a distinct bias towards carbon-12, because this lighter isotope is more mobile.

Some of Viking's nutrients, along with small reservoirs of carbon dioxide and carbon monoxide gases, were laced with a third variety, carbon-14, a heavier isotope than one would expect to find in natural organic chemistry—and also radioactive. With miniature Geiger counters, the lab's instruments would be able to determine whose carbon they were sampling at any given moment: ours or theirs....

Now to the actual experiments, and what they found. The first cycle of tests at the Chryse Planitia landing site was initiated on July 28, 1976, when Viking I extended its robot arm and scooped up a fistful of soil. This was not a process that could be steered in real-time from earth. As we have observed, the machine had to carry out complicated procedures without human guidance because of the 19-minute radio time lag. Pasadena had to depend on Viking's computers to bring in the soil and distribute it fairly between the various compartments of the laboratory box. Then all the experiments had to run simultaneously, because certain pieces of the general hardware, such as heaters, gas pumps and data tape recorders, were jointly shared between the different experiments.

Does anything in Martian soil exchange gases with the atmosphere?
THE GAS EXCHANGE EXPERIMENT (GEX)

The Gas Exchange Experiment's team was headed by Vance Oyama from Nasa's Ames Research Center in California. The GEX assembly detected any transfer of gases, organic or inorganic, between the soil and the surrounding atmosphere.

Nutrients were added to the sample in order to stimulate a reaction, but only very gradually. There was a good reason for this. As we have discussed in connection with Vishniac's Wolf

Trap, there was a feeling that simply drenching a soil sample in water-dissolved nutrients might be counterproductive. A graduated approach allowed for minimal contact at first, with the soil initially exposed to a very fine mist of nutrients, but not actually soaked. The soil was suspended in the middle of the chamber in a little porous cup. This was called the 'humid mode'. Some days later the sample was submitted to a 'wet mode', which allowed an actual fluid contact with the soil.

Throughout the test, the atmosphere around the soil was constantly monitored for any changes in composition. The measurements were sustained over several weeks. It should be noted that a little cheating was required to prevent the nutrient from freezing or evaporating in the natural conditions prevalent on Mars. The test chamber was sealed, and gently warmed.

At periodic intervals, the atmospheric mixture above the soil was extracted by pumping in helium. This chemically inert gas would not distort any of the test results. The expelled mix now passed through a 'gas chromatograph', a column containing a granulated filter which impeded the mix's progress from one end of the column to the other. Certain types of molecules would pass through very fast, others more slowly. Just as a drop of ink seeps across a sheet of blotting paper and separates out into its different constituents, so it was with the GEX's expelled mix of gases. They came out the other end of the column at different rates, separated into their individual types. The chromatograph was suited to identify fairly simple substances like hydrogen, oxygen, carbon dioxide, nitrogen, and the simplest hydrocarbon, methane (CH_4), all of which can be found among the range of microbial waste products on earth.

The GEX was among the first of Viking's experiments to send back data from the surface of Chryse Planitia. The results electrified the waiting scientists on earth. After less than three hours in 'humid mode' the soil gave off a tremendous burst of oxygen, far greater than anyone had anticipated. At peak levels, there was 200 times as much oxygen in the atmosphere above the sample than had been present at the start. Bear in mind that this ratio was compared with the typical composition of the atmosphere, which usually contains negligible amounts of free

THE EXPLORATION OF MARS

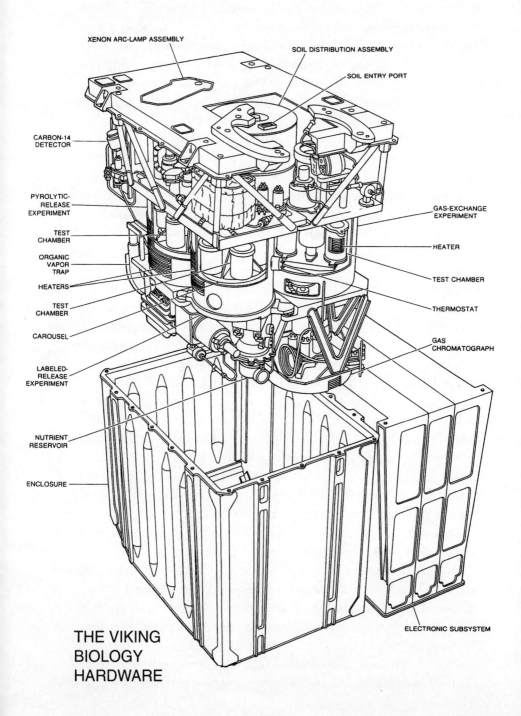

THE VIKING BIOLOGY HARDWARE

oxygen, about a tenth of one per cent. The 200-fold oxygen increase up to 20 per cent was certainly startling. However, the marked rise in the GEX gas levels as a whole was mainly a result of another product, carbon dioxide, which also emerged from the soil, boosting the overall pressure in the chamber by a factor of five. Viking scientist Dr Gilbert Levin said at the time, 'Never have we seen that magnitude of response in any of the earth samples we tested that did not contain life.'

Eventually, after about 50 hours, the oxygen release tailed off. Carbon dioxide levels climbed on for a while, but much more slowly than in the first few hours. If something in the Martian soil were alive, presumably it would continue working for as long as fresh nutrients were available. So, why had the reactions ceased while there was still plenty of nutrient to feed from? More ambiguities emerged from data relating to the speed at which the initial reaction had taken hold. The rise in activity seemed too sharp to have been biological.

After 16 Martian days, the soil was subjected to the 'wet mode'. Carbon dioxide levels showed slight signs of increase, but the oxygen release had largely ceased. By day 25, everything had levelled out for good. It looked as if some non-living chemical reaction had worn itself out during the earlier 'humid mode'. So where had that powerful initial burst of gases come from? Thoughts turned to the behaviour of substances like hydrogen peroxide, H_2O_2, which readily decomposes in the presence of metals and water, liberating oxygen. To the proponents of a non-living explanation for the GEX data, this seemed a likely answer. Mars was obviously very rich in iron oxides. Perhaps, over millions of years, native peroxide compounds had attacked all the metallic elements of the Martian dust and soil to produce those oxides? Maybe just the *water* in the nutrients had triggered the GEX reaction, liberating oxygen from peroxides? In which case, it was as if Viking had delivered nutrient toys to Mars only for the planet to show more interest in the watery wrapping paper than the actual gifts.

However, the 'pro-lifers' pointed out that peroxides are very unstable substances. The whole point about such chemicals is that they don't maintain their integrity over time. Hydrogen peroxide can be thought of as a water molecule but with two oxygen atoms bound to the hydrogen where only one should be. Hence its formula, H_2O_2, as opposed to the more familiar H_2O. The pairing between the two oxygen atoms is weaker than the links in water between oxygen and hydrogen, and it is very easily disrupted.

In fact, hydrogen peroxide is so unstable that it can be used as a rocket fuel which, in essence, burns itself without needing an independent oxidizer. As some engineers have discovered to their cost, you don't always need an ignition flame to start off a reaction. Hydrogen peroxide will often explode spontaneously. On Mars, it would only take exposure to sunlight for peroxides to decompose. How could such compounds remain on Mars for any significant time?

To deepen the mystery, Mariner IX's spectroscopes hadn't detected peroxides on the Martian surface, nor in the atmosphere. Viking's orbiting mothership confirmed this absence. Sophisticated (i.e., cleverly speculative) theoretical models for the temporary formation of peroxides in the Martian environment showed that the stuff should be destroyed a hundred times faster than it is created. So the dramatic release of oxygen in the GEX experiment was not adequately explained by these models, because peroxides had absolutely no business being in the soil in the first place.

As for the release of carbon dioxide: this gas is the principal constituent of Martian air, and it would have saturated every speck of soil. When you stick two polished surfaces of different material face-to-face, weak electrical forces at the atomic scale often hold them together without actually creating genuine chemical bonds. A process called 'adsorption' performs similar tricks with gases swirling into the tiniest pores of certain types of mineral surfaces. It's as if one polished surface bends with infinite flexibility to meet another. Maybe the water from Viking's nutrients was simply displacing the adsorbed carbon dioxide from the soil?

At least this explanation had the benefit of not relying on peroxide theories for a simple solution—though it didn't account for the burst of oxygen, which was clearly a chemical product rather than a simple physical translocation. Either the oxygen was being liberated from all those oxides (and maybe peroxides) in the soil, or else it was coming out of microbial metabolism. 'Free' oxygen could not have persisted in the soil for long. This applied equally to the strange burst of oxygen observed by the GEX instrumentation. If it was indeed being released from a peroxide in the soil, then in theory it would be free to attack the organic compounds in the Viking nutrients, breaking them down and oxidizing carbon atoms. In that case, the observed CO_2 wasn't purely a product of something deep in the soil reacting with the nutrients—it was just a misleading by-product of the oxygen release.

As the GEX debate deepened, one of the Viking scientists, Gilbert Levin, commented, 'So far the score is Biology, one, Chemistry, one.'

Then the other experiments began their test cycles. Levin would shortly amend his assessment of the score.

Does anything in the soil release carbon?
THE LABELLED RELEASE EXPERIMENT (LR)

In 1967 Dr Gilbert V. Levin founded Biospherics Incorporated, a small commercial company in the environmental testing business. Viking's Labelled Release Experiment (LR) was a clever adaptation of Levin's patented method for detecting potentially dangerous bacterial contamination in water supplies, foodstuffs and hospital blood banks. At a cocktail party some years earlier, Levin had chatted briefly with a Nasa administrator, and he immediately realized that his detector could be adapted for a more exciting function than bug-hunting among the nation's public utilities and hamburger factories. It could just as well seek life on other planets.

The LR searched for carbon-based gases released during metabolism. A soil sample was placed in a container, along with a volume of uncontaminated Martian atmosphere. Then the soil was humidified with a fine mist of nutrients laced with carbon-14. Throughout several days of gentle incubation, the Geiger counters looked for carbon-14 atoms in the atmosphere around the soil. If any were detected, it could be assumed that something in the soil was taking up carbon from the nutrients and then expelling it as part of a waste gas.

Levin argues to this day that his LR experiment yielded the most positive of all Viking's biology results. As soon as nutrients were introduced, the Geiger counters clicked contentedly as atom after atom of radioactive carbon was given off by the soil, producing 9000 blips per minute for seven Martian days. The single-count blips indicated the atoms' presence in a simple compound, which was almost certainly carbon dioxide (CO_2). Of course the Geiger counters could hunt only for carbon-14 atoms. They couldn't determine the kind of molecules in which they might be bound up.

Theoretically, single blips could have been registered by methane molecules (CH_4). On present-day earth, methane is entirely the product of the anaerobic decomposition of organic material. The discovery of methane emerging from Martian soil would have been tremendously exciting, but unfortunately, the Viking data never came close to suggesting this possibility.

After seven days the LR carbon counts fell away. Levin and his co-worker, Patricia Straat, wanted to leave the experiment alone indefinitely, to see if the count climbed up again. They sought to eliminate the possibility that a 'one-off' reaction had simply burned itself out. As we have noted, the experimental cycles were determined by hardware considerations as well as by scientific requirements. The other Viking scientists were keen to initiate different parts of their own procedures, and would not support a long hold-up while the LR chugged on and on. Just as Straat thought she might be detecting an interesting statistical anomaly in the supposedly quiescent LR carbon count, she and Levin had to end that particular cycle.

Levin now lobbied intensively for a chance to run a 200-day LR test, either at Chryse Planitia or later, when Viking II was safely down at another site. This would have meant locking all the *other* experiments into a 200-day waiting period. Levin met a great deal of opposition when he made that particular request. Every day that went by, Mars was speeding in its orbit towards 'occultation', the point where it would disappear behind the sun, blocking out the radio link with Viking. That event was scheduled for early November 1976, just three months away. Nobody but Levin wanted to waste 200 days out of the time remaining, just in case Viking failed to make contact when it came round the other side of the sun in December.

Levin suggested that his test could run *while* the Viking was out of contact. The accumulating data could be stored on a tape recorder and then transmitted to earth once the radio link was regained. The consensus among Viking's engineers was that all the hardware should be put into a dormant condition, just as planned many months earlier.

On this occasion Levin yielded to the inevitable and ran the next cycle of his experiment in lockstep with everybody else. The LR canister had soil samples in reserve, and a completely fresh batch was swivelled into position. This time, though, the nutrients were withheld for several hours while the sample was 'pyrolyzed', baked in a mini oven to 170 degrees Celsius. This temperature was high enough to reduce any bugs into their constituent molecules and destroy them, but too low to break down the more temperature-resistant silicates and iron compounds in the bulk of the soil.

The key point is that Levin turned on the heaters *before* the nutrients were added. After that, the Geiger counters yielded no carbon blips at all throughout a ten-day period of monitoring. Something in the Martian soil had shown precisely the same kind of sensitivity to heat that one would expect of organisms.

Unfortunately, peroxides also boil off at 170 degrees Celsius. If they were somehow present in the soil, their reactiveness would also have been eliminated by pre-pyrolysis. Could peroxides have produced the earlier carbon releases? Levin didn't think so. The GEX may, in theory, have simply liberated existing atmospheric carbon dioxide in the soil through desorption (the opposite of adsorption), but the LR had shown the soil taking up carbon atoms from Viking's organic nutrients and then expelling them in newly synthesized carbon dioxide: a classic incidence of the carbon 'cycle' which characterizes life. *Old Martian carbon dioxide obviously couldn't have included Viking's radioactive isotopes.* As Levin says today, 'As far as the LR is concerned, we got data satisfying the pre-mission criteria for the presence of life.'

The counter-argument from the anti-lifers was the same as for the GEX experiment. According to them, Levin's freshly-minted CO_2 was a product of oxygen released from peroxides which then attacked his nutrients. Once again, it seemed quite plausible that the gas was not actually created *inside* the soil.

Levin wasn't having any of it, though. In defence of his LR results, he became adept at turning the peroxide supporters' own arguments against them. If a peroxide was liberating oxygen, and if it was attacking Levin's nutrients, then according to the GEX data there should be more than enough oxygen in any given soil sample to attack most, not just some, of the labelled carbon in the nutrient. Levin's carbon counts certainly yielded a strong result, but it was only one seventh as strong as it *would* have been if oxygen had attacked all the nutrient's carbon without bias. By a neat coincidence, Levin's mixture of nutrients contained seven varieties of organic compounds. As far as he was concerned, only a seventh of the available carbon atoms had been assimilated into the new CO_2, and therefore the soil must have shown a bias for one of the nutrients and ignored all the others. Raw oxygen would have attacked all seven compounds indiscriminately. Levin's critics said that the one-seventh ratios were just coincidental.

Does anything in the soil assimilate carbon?
THE PYROLYTIC RELEASE EXPERIMENT (PR)

The PR experiment checked to see if Martian soil assimilated carbon from the atmosphere. First, a portion of soil was placed in a thumb-sized container. Then, a measured mix of radioactively labelled carbon dioxide and monoxide gases from Viking's reservoirs was pumped in, and the chamber was sealed. A small xenon arc-lamp simulated Martian sunshine, minus the harmful ultraviolet radiation, which was blocked out by a quartz transparent filter.

This was a bit of a cheat. One had to assume that there would be no such privileged protection against ultraviolet at the shallow soil levels where Viking's scoop could reach. The filter was designed to prevent certain peculiarities in the test results rather than to protect any Martian bugs. In pre-flight experiments Horowitz and his colleague Jerry Hubbard discovered that ultraviolet radiation acting on radioactively labelled carbon monoxide and traces of water vapour produced small quantities of organic compounds on certain types of soil surface; in particular, formaldehyde (HCHO). That was a fascinating result in itself, but it was not desirable that Viking's experiments should be capable of manufacturing their own organic compounds with very little help from Mars.

Anyway, the soil was exposed for five days to the lamp's radiation. At the end of this incubation period, the surrounding gases were flushed out of the chamber, once again by pumping in helium. Any free-flying traces of carbon-14 were thus removed from the experiment. Now the soil was subjected to pyrolysis, the heat treatment designed to decompose any Martian organisms. This time the intention was not merely to kill them off, but to extract their constituent molecules as vapours which could be sampled by the Geiger counters for radioactive carbon-14 atoms. If any were detected, it could be assumed that the microbes had assimilated some of Viking's labelled carbon into their internal chemistries.

While the LR experiment had searched for carbon coming out of the soil, the PR looked for it going in. There's not necessarily any contradiction here. One can think of the different tests catching carbon at different stages in its 'cycle'. Imagine that you ate a sandwich with labelled carbon in the margarine. A breath test 24 hours later would provide a radioactive count as you exhaled CO_2 molecules. Many of these would include some of the margarine's original carbon.

At the same time, other labelled carbon atoms would have been taken up to build new body tissues. They would be easily detectable, should you ever choose to surrender yourself to pyrolysis inside a Viking robot.

In the PR procedure, however, no nutrients were added to the experiment, and the focus was on a 'plant'-like chemistry whereby Martian microbes might take all they required from the existing soil, the atmosphere and the sunlight without any need for extra food. In essence, Viking added nothing the bugs wouldn't already obtain from their native environment, save for the radioactive labelling of carbon.

The mirror-image similarity between the LR and PR was neat, but it was more a fortuitous product of bureaucracy than of careful advance planning.

When Levin first took up his Viking assignment, he had not quite satisfied Nasa that his scientific qualifications were up to scratch. His candidacy for inclusion in the project was based in part on the solid track record of his company, Biospherics, and on the obvious value of his ideas. However, Nasa's senior managers were wary about issues of academic respectability. They requested that Levin take on a co-worker, Norman Horowitz from the California Institute of Technology, who had all the right credentials. Levin went on to acquire his own PhD, but if anything, Horowitz—the PR experiment's creator and team leader—seems to have learned from Levin on this particular occasion rather than the other way around. Hence the PR's cousinship with the LR.

The PR's results were not so dramatic as the LR's, but they were intriguing nevertheless. The amount of 'fixed' radioactive carbon detected in the soil was marginal, but it was there. The best estimates suggested a quantity of carbon which might have been assimilated by between 100 and 1000 bacterial cells. This would represent an incredibly tiny physical volume of biological mass; but *no* organisms, not even the smallest quantity, should have survived in soil steeped, say, in hydrogen peroxide, which is highly destructive to organic chemistry. Horowitz, a deeply cautious scientist, had originally predicted the odds of finding any life on Mars as less than one in two hundred. He had not expected a positive result from the PR. 'You could have knocked me down with one of those Martian pebbles,' he said afterwards. Along with the LR data, the PR results militated against the peroxide theory and seemed to give the pro-lifers an edge.

There were soil samples in reserve for the PR test. A fresh batch was swung into position. This time, instead of pyrolyzing the soil after the experiment was over, the heaters were switched on before the test, to see if anything alive in the soil might be affected by heat—the 'control' procedure. Apparently something *was* badly damaged, if not entirely eradicated. When the radioactively labelled gases were added to this pre-pyrolysed soil, the subsequent carbon take-up was barely 10 per cent as strong as before.

Life would be killed off at the 170-degree temperature of pyrolysis, as we have observed. Any microorganisms would indeed have been broken apart at that temperature. But then again, pyrolysis over several hours should have ruthlessly killed off all the bugs, not just 90 per cent of them. For that matter, it should also have destroyed any hydrogen peroxide in the soil.

The data was confusing. The low number of carbon blips registered by the Geiger counters hovered tantalizingly on the edge of the PR machine's tolerances. If the count could be attributed to random noise in the hardware, or to natural levels of background radiation, it might be possible for Horowitz to conclude that the pyrolysis had completely eradicated the soil's ability to react with Viking's gases, or nutrients, just as in Levin's LR tests.

Horowitz was emboldened by an important consideration. Because the PR experiment did not involve nutrients, but used dry gases instead, there was no water to muddle the results or to precipitate strange oxidation reactions in the supposed hydrogen peroxide in the soil. Horowitz had consistently opposed carrying liquid nutrients aboard the Viking. He had championed the opposition against Vishniac's Wolf Trap, for instance, insisting that microorganisms on Mars, if they existed, would be so extremely adapted to ultra-arid conditions that the merest contact with water vapour in the 'humid mode', let alone a good soaking in the 'wet mode', might be enough to burst them apart. Horowitz dismissed all the nutrients, devised after years of research, as 'chicken soup'.

Oyama's GEX team resented the implied criticism of their procedure. Horowitz was barely able to conceal his 'told-you-so' expression when the peroxide-based theories to explain the GEX data fell prey to the same kind of water-based confusions he had always warned against. He knew that if anything should happen during PR, it could not simply be passed off as a purely accidental product of water and peroxide, because he wasn't using any water. 'My experiment is the only one that's actually like Mars', he said at the time.

With this thought in mind, Horowitz tried to replicate his first experiment, the one which had observed a level of carbon 'fixing' equivalent to a thousand microbes. He wanted to confirm that something had happened purely spontaneously in the soil. The surest way of achieving this would be to obtain the same kind of results from a fresh soil sample.

Unfortunately, just as the PR was supposed to show off its freedom from potential confusions, it generated a muddle of its own. The pipework had sprung a small but persistent leak. Horowitz calculated that the radioactively labelled gas was escaping from the test chamber at the rate of about 6 per cent each day during a run. Since Horowitz needed to incubate his sample in contact with the gas for at least five days, his carbon counts at the end of that time threatened to be out by 27 per cent. From now on, he would have to compensate by adjusting all his new data in order to boost the carbon counts artificially. This would yield the sort of reading he *might* have obtained if the volume of labelled gas in the experiment had been up to its proper strength for the whole time, instead of simply pouring out of a tiny hole in the pipework. Needless to say, this contingency greatly increased the margin for error. The more the gas poured away, the more Horowitz had to fudge his figures. Now, Oyama and Levin could tear apart Horowitz's findings as easily as he had dismissed theirs.

Although the initial PR result had been exciting, Horowitz was never particularly 'pro-life' in his expectations. He put the odds at less than a hundred to one. If anything, he now wanted to clarify his results to support his growing suspicion that peroxides lay at the heart of the entire mystery. At a press conference on August 7, 1976, to discuss the first cycle of experiments, the ambiguity of his position was matched by the twists and evasions in his statement. 'I want to emphasize that we have not found life on Mars', he said. 'However the data points we've obtained are conceivably of a biological origin, but biology is only one of a number of alternative explanations that have to be excluded.

THE VIKING INVASION

We hope to have excluded all but one of the explanations, whichever that may be.'

Some journalists weren't too sure if that was a 'yes' or a 'no'.

Does the soil contain its own organic compounds?
THE GAS CHROMATOGRAPH/MASS SPECTROMETER

The Gas Chromatograph/Mass Spectrometer (GC/MS) team was headed by Klaus Biemann from the Massachusetts Institute of Technology. Their experiment was supposedly the most critical. The hardware was complicated, with its own canister completely distinct from the GEX, LR and PR assemblies, its own privileged supply of soil, and a device for grinding the sample down into tiny fragments that could easily be vaporized.

On July 28, when the soil scoop was pulled in towards the main body of the Viking craft, the biology box with the GEX, LR and PR experiments inside it received the first portion of soil. Then the arm reached out to grab another sample for the GC/MS. It was desirable that everybody should work with similar soil, so the arm went back to the trench it had dug earlier.

In all the simulations of digging which Nasa had carried out on earth, all the freshly-dug trenches in dry, fine soil collapsed in on themselves to some extent, so that second digs at the same location were always possible. On Mars, however, the super-dry conditions produced a different phenomenon. The walls of the first trench remained crisp and firm, and no collapse took place. Consequently, when the robot arm tried for a second scoop, it just scrabbled at empty air. When it swivelled around to the GC/MS box destined to receive its new load, Biemann's team watched their consoles at Pasadena, awaiting the signal that would confirm the actual presence of soil in their experiment. The indications weren't quite clear.

Unfortunately the test cycle was due to be initiated. For all anybody knew, the $55 million GC/MS might even now be preparing to analyse the contents of an empty chamber. 'I'm on the non-receiving end of the system,' Biemann complained.

A photograph of the trench confirmed what everybody had feared. Its flanks were as steep and smooth as concrete walls. On the other hand, a photo of the top of the GC/MS box showed a scattering of spilled grains, so some soil had probably reached it after all. The box was supposed to be able to test a total of three samples during the mission, but one of its 'dolls-within-dolls' was broken. Biemann had just two test opportunities available to him. He wasn't anxious to waste half of them on sampling an empty cell.

Then the robot arm started playing up. For an anxious day or so, it seemed that the scoop might never be able to reach the ground again to fetch more samples. Crossing his fingers and praying that the GC/MS had somehow managed to swallow at least a smattering of soil earlier on, Biemann ran the first cycle in the knowledge that it might be his only chance, at least as far as Viking 1 was concerned. The second lander hadn't yet touched down, let alone reported its experiments fit for duty. What if it crashed? If he turned down this current opportunity to do a run, Biemann might end up junking the entire GC/MS effort. It was best to cut his losses and go for broke.

Of course, just as he had committed himself to a possibly flawed run, the Nasa engineers figured a way of getting the damned arm to work again. Apparently the mechanism didn't like the early morning cold. It was more willing to perform in the mid-afternoon, when the sun climbed higher in the sky and the prevailing chill became slightly less extreme. Biemann 'cursed all robots everywhere'.

For the GC/MS, Viking's nutrients were withheld so as not to confuse the final analysis. The chamber was heated to 500 degrees Celsius, to break down any native Martian organic compounds into vapours which the GC/MS could detect. The first part of the assembly, a gas chromatograph (GC), was similar to the granular column in the GEX experiment; but a chromatograph on its own is best suited to identifying fairly simple gases—hydrogen, oxygen, carbon dioxide and methane for instance, as we have noted. The chromatographic filter would still serve to separate out more complex organic molecules, but their identification would require an additional procedure.

For a start, the temperature in the soil chamber was raised in carefully measured stages, with six-minute pauses between each climb. In this way, different 'families' of organic compounds might be driven from the soil at discrete intervals, so that they could be analysed separately. Once the vapours emerged through the chromatograph, supposedly separated into their different categories, they were pumped past a small beam of charged particles, fired from the same kind of device one might find in the back of a TV tube. This bombardment ionized the passing gases, imparting a positive electrical charge to their molecules by dislodging electrons. The final part of the GC/MS assembly, the mass spectrometer (MS), could now measure to what degree the ionized gas molecules were deflected in their path by a magnetic field, with heavier molecules being less affected as they hurtled past than lighter ones. The molecules would finally slam into a screen of electronic detectors, hitting different positions on the screen according to how far, or how little, they had been deflected. Benchmark tests of various organically-derived vapours on earth could then be compared with all the GC/MS data from Mars.

The GC/MS picked up a whiff of cleaning solvent left over from its own manufacture back at the TRW factory on earth. At least the machine was working. Apart from that, though, it found nothing but carbon dioxide and a trace of water vapour in the gases driven off by the hot soil. All the remaining substances in the soil were still back in the test chamber: the iron oxides, silicates and so forth, which would have needed at least twice as much heat to vaporize. Biemann and his team felt fairly confident that at least some soil must have got into the GC/MS chamber after all.

In theory, their results *could* have been derived from a chamber containing nothing more than Martian atmosphere, but any such gases had supposedly been flushed out of the chamber before the run. The other possibility was that the suspected hydrogen peroxide had decayed under heating, liberating water molecules. If this were the case, then why had the GC/MS not picked up the excess oxygen in the peroxide, which should also have been freed by the heat?

For the time being, most of the Viking scientists were ready to admit that the GC/MS had worked more or less properly. So, if the LR and PR tests had found apparent evidence for microbial metabolisms earlier, the big question now was: why hadn't the GC/MS test sniffed the bugs' heat-shattered molecules in its own soil sample? A population of microbes, no matter how small, is not just the sum total of its active individuals. There is usually an adjacent clutter of dead material. A thousand living microbes should have been part and parcel of a larger quantity of organic matter. The failure of the GC/MS to detect any of this was a serious argument against the presence of living entities.

To his credit, Biemann proposed a possible explanation, even though his theory impugned his own instrument, the GC/MS: 'There may have been too small an amount of organic material for it to detect.' Biemann suggested that conditions on Mars were now so harsh that microorganisms might have evolved to be as efficient as possible, reabsorbing dead material and shedding very little of themselves while alive. If the Martian organisms were in deep hibernation, this might also account for the very small population, with just a handful of hardy survivors remaining from better times long ago.

It was risky enough to make the assumption that life on Mars was chemically similar to life on earth. Were the scientists also expected to double-guess the details of an alien evolution and lifestyle?

'No', says Gilbert Levin today, 'there wasn't a good argument to say that a population of a thousand microbes in the sample was unrealistic. You would expect Martian organisms to be very thinly distributed, especially if semi-dormant, and they would be very efficient at reabsorbing their own spent chemistry. All the wasteful organisms would have died out long ago. Life can be incredibly hardy, we already know this. It's difficult to kill off completely, and if it ever got going on Mars, I think it should still be there today.'

CHAPTER THREE
NORTHERN EXPOSURE

THE SECOND SPACECRAFT, Viking II, achieved Martian orbit on August 7, 1976. Its orbit was inclined slightly differently to that of the first ship, opening up a tranche of landing options in the upper northern hemisphere. The intended landing site was Cydonia, a plain located about 1500 kilometres north of Chryse Planitia, near the edge of the ice cap.

In mid-1976, Mars' northern hemisphere was in the middle of its six-month summer, and the ice was retreating. There was an excellent chance that the atmosphere at Cydonia would be rich in water vapour, and the ambient temperatures were expected to rise a few degrees above freezing in mid-afternoon. The plain as a whole incorporated a gentle dip, so that at its lowest, the ground was a good 20,000 metres below the mean surface level of the planet. This would have a marked effect on atmospheric pressure at ground level; not so marked as the 7-kilometre depths of Marineris, but perhaps enough to make a difference. There was a slight but tantalizing chance of finding traces of liquid water in the soil.

Nobody was entirely surprised when Cydonia turned out, on closer inspection from orbit, to be less friendly than expected. The biology team lobbied for the lander to be sent even further north, to the very outermost edges of the ice cap itself. They reasoned that permafrost in the soil might briefly become liquid in summer, and Viking's chance of detecting water would be that much better. However, this nudging northwards of the touchdown target was a tricky game to play. If the lander came to rest anywhere above the 55th parallel, it would work well enough during summer, but the winter in six months' time would be too cold for the machine to withstand. Even so, the argument for a northerly site was supported by the orbiting mothership's analysis of the atmospheric water vapour content. In certain areas this was ten times greater in concentration than at Chryse.

The permafrost argument certainly made sense. A great number of Martian craters show undulating patterns around their rims that seem strangely rounded, like frozen mud flows. The central impact scars are also somewhat mushy. If meteorites slammed into a mixture of permafrost and loose rubble, the heat of the impact would churn such a surface into viscous waves of slurry. If the surface refroze fast enough, the viscous slurry's strange, petal-like waves might be held in place for ever. Wherever such a crater could be found on the Martian surface, the presence of permafrost could be assumed. Those who supported the idea that Mars might once have been a warmer and wetter world firmly believed that most of the ancient water would be found just underneath the planet's dusty rubble-strewn surface.

Most of the craters at Cydonia turned out to be even stranger. They were inside out, sticking upwards from the surface instead of digging downwards. The heat and pressure of the original impacts must have compressed a fairly loose surface. This would explain how, over countless millennia, the Martian winds tore away at Cydonia's weak topsoils, carving out the plain's gentle dip, while the toughened crater floors resisted the long barrage of erosion.

A story from the world of show business illustrates this phenomenon nicely. The TV mogul Lew Grade once filmed an expensive and unsuccessful project called *Raise the Titanic*. He commented, bitterly, that it would have been cheaper to lower the Atlantic. The inverted Cydonia craters were a product of just such an extraordinary circumstance. They were a fascinating geological discovery, and of course, a daunting threat to Viking II's soft underbelly. The descent onto Cydonia was cancelled.

On September 3, the lander at last dropped down onto Utopia, a flat plain almost exactly halfway around Mars from the Chryse site. The biologists had nearly got their wish. Utopia was on the 50th parallel, five degrees further north than Cydonia. To everybody's relief, the lander touched down intact.

All things considered, the mission as a whole was proceeding very well. There were now four vehicles on or in the vicinity of Mars. The two landers relayed their signals to the two orbiters, which in turn sped their signals towards earth. Any one of the quartet of ships could communicate independently if required.

Although the general public thought of Viking as a single machine, Pasadena was managing the activities of a small fleet. The fact that the two landers were safely settled on the surface made little difference. They were still in motion as the planet spun around on its axis. The earth, the landers and the orbiters constantly shifted about in relation to each other in a pattern of interlinking curves that never quite repeated themselves.

While the Nasa space technicians patted themselves on the back, Viking's biologists had yet to discover whether or not the new landing zone would answer any more questions for them. When the first pictures came back, the terrain was revealed as even more littered with rocks and boulders than Chryse. It looked pretty unforgiving. Utopia wasn't exactly Utopian.

Once more, the biology team geared up to run their experiments. They wanted to cross-check their results from Chryse, and also to see if Utopia would reveal any dramatic differences in its composition and reactivity. Biemann, of course, wanted to make sure he got a decent scoopful of soil for the GC/MS this time around. He did, and the Utopia results came out broadly similar to Chryse's, with no trace of organic compounds.

Meanwhile Horowitz ran his PR again, and consistently failed to get a carbon 'fix' rating as high as his first attempt on Viking I. However, he did obtain readings more than twice the level at which they might be discounted as background noise in the instrument. The peroxide theorists were not surprised by the PR's failure to peak. Utopia was 'wetter' than Chryse. Any peroxides in the soil would have been depleted by reactions with water vapour (catalyzed by the orange-red iron oxides) during the summer thaw.

Vance Oyama's GEX team got an oxygen burst, and he was delighted with this proof of the validity of his startling results from Chryse. Even so, this time the burst was only a quarter as intense as before. Again, the peroxide argument suggested a partial burn-out of reactivity in the presence of summertime water vapour in the Utopian soil.

Predictably, Levin and his LR experiment bucked the trend. The labelled carbon dioxide given off by Utopia's soil in the presence of nutrients produced a count 25 per cent *stronger* than at Chryse, while every one else's experiments showed weaker responses. Triumphantly, Levin argued that if peroxides were responsible for all the reactions in the GEX, LR and PR experiments, then there should be some proportional mathematical relationship between the three sets of results. His LR seemed to be out on its own, chugging nicely away against the odds. Levin's critics pointed out that the LR carbon counts still tailed off eventually, just as they had on Chryse.

Remember that on Chryse Levin had pyrolyzed a second soil sample to kill off any biological reaction. This had done the trick. Now he wanted to discredit the peroxide argument once and for all by pyrolyzing again, but this time at a much lower temperature, on the very cusp of the heat levels required to neutralize any bacteria. Levin reasoned that the new temperature, 50 degrees Celsius, should be high enough to knock a bug's metabolism out of kilter by damaging enzymes and tearing internal structures apart. However, the heat should not be great enough to vaporize them altogether. Crucially, at such a low temperature, no inorganic compounds were likely to be affected at all; not even the wretched peroxides.

Levin duly ran his pyrolysis at a lowly 50 degrees. He nicknamed the procedure 'cold sterilization'. He had to run this cycle a couple of times because of instrument anomalies, but he got the result he was looking for. At 50 degrees his Utopia soil lost all its reactivity. At this point the peroxide theorists, led by Horowitz, suggested that, in the low atmospheric pressure, Martian peroxide could have boiled off at a lower temperature than it did on earth. At that, Levin exploded: 'You can't just keep moving the goalposts!' he protested. He quoted textbooks at his rivals, and reminded them that the pressures inside all the Viking test chambers were in any case artificially higher than outside. Otherwise the nutrients would freeze or evaporate. Levin then found a respected scientific paper from Germany which indicated that peroxides might indeed have decayed at 50 degrees in a low-pressure environment, but only at the rate of 3 per cent an hour. Since his 50-degree pyrolysis at Utopia had lasted only three hours, any peroxides present in the sample shouldn't have lost more than 9 per cent of their strength.

Levin refused to allow the case for peroxides to prevail. 'There is not one jot of direct evidence that they exist. I don't believe in them,' he maintained. It was inconceivable, he said, that any inorganic chemicals would have decayed at such a low temperature. It must have been *organic* compounds which were disrupted by pyrolysis. Therefore, organic compounds in the earlier non-pyrolyzed samples must have been responsible for assimilating radioactive carbon from Levin's nutrients and releasing it in fresh-minted carbon dioxide. Life on Mars. QED.

In 1976 Levin was prepared to say only that 'I'm leaving the door open to the biological explanation', but among his colleagues there was no doubting where his true beliefs lay.

Norman Horowitz found himself in an awkward position. His interpretation of the Viking data was at odds with Levin's, but he was unable to clinch his side of the argument without resorting to speculation. Every scrap of his experience told him that all the reactions came down to inorganic processes, and that some kind of exotic oxidizing agents existed in the Martian soil. The trouble was, he couldn't prove it.

For his part, Vance Oyama had also become convinced that the GEX had recorded an inorganic reaction. Wary of the many inconsistencies produced by Levin's data, Oyama modelled a complex chemistry in which more than one kind of compound was at work in the soil. He posited hydrogen peroxide in close association with 'superoxides', which are even more unstable substances based around metallic oxides but with the usual top-heavy oxygen content. Horowitz backed him up. 'In the absence of living organic processes to mitigate them, why shouldn't Martian soil contain a number of different oxides?'

Levin described these ideas as 'conjectural'. It was about the worst insult he could think of throwing at his fellow scientists. By this time, the biology team was fractured from top to bottom. Chief biologist Harold Klein tried to mediate. 'The lander's experiments are like four guys in a fishing boat. A couple of them have felt nibbles on their lines. Right now we don't know if they've caught a fish or snagged a beer can.'

A SIMULATION OF LIFE?

Carl Sagan, usually the most ebullient of all the Viking scientists, had been curiously subdued since the first pictures came back from Utopia, and at first contributed relatively little to all the arguments. True, his principal role was to serve as a member of the imaging team, but his previous track record in planetary life sciences entitled him to enter the biology debate if he so chose.

It was obviously a disappointment to Sagan that the two landing sites, separated by a whole planet's width, one of them broadly equatorial and the other in high northern latitudes, had turned out to be so similar.

Suppose a pair of Viking landers had touched down on earth at equivalently separated places—say, a Nebraska cornfield and a patch of Icelandic tundra. There would be many shared chemical characteristics, but some noticeable disparities would have emerged. For instance, Nebraska farmland would have proved greatly richer than Icelandic tundra in nitrogen and phosphates. No such obvious dramatic differences had been revealed by the Martian sites. Indeed the whole planet seemed a poignant contrast to earth's liveliness, as Sagan sadly pointed out in a litany of despair. 'There are no bushes, no trees, no cacti; no giraffes, antelopes or rabbits; no burrows, tracks or footprints. There are no patches of colour that might be photosynthetic plant pigments .... So far, none of those rocks has obviously got up and walked away.'

Slowly Sagan's mood improved, and at last it dawned on him that it didn't really *matter* if Viking had failed to find conclusive proof of life. 'Four instruments were designed to detect life, and three of them produced positive signals. What does this mean? Even if it turns out that the results are non-biological, then there are obviously chemicals that simulate the behaviour of life just lying around in the soil.'

Sagan recognized that finding the possible precursors of biology could be just as momentous as discovering actual living entities. For there was one thing that most of the Viking scientists could agree on: earth and Mars were originally shaped from the same solar system stock, and as far as anyone could see, there was no particular reason why Mars had to have lost out in the great game of life. Mars might reveal what the earth was like at a time when plants and animals had not yet scattered their kind throughout its oceans and across virtually every scrap of land. If we could find out why Mars failed to produce life—if, indeed, it did fail—then we might unlock valuable clues about the earth's earliest ecology before the all-encompassing emergence of full-fledged organic biology. This is a point we will return to during the course of the next chapter.

MEANWHILE ...

Nasa's Martian geologists did not receive so much attention from the world's press as their colleagues in the biology team. Nevertheless, they set about their tasks with equal enthusiasm. They, at least, got some idea of what Mars was made of: oxides of iron and titanium, silicon dioxide, aluminium, magnesium, calcium and sulphur. The iron oxides in particular give the planet with its distinctive red colouring.

Even this data wasn't free of ambiguities. The geologists had their own instrument, distinct from all those of the biology tests. Their 'X-Ray Spectrometer' bombarded fragments of soil with X-rays (electromagnetic radiation of very short wavelength) so that atoms on the fragments' outer surfaces would fluoresce, generating their own X-rays at varying wavelengths, depending on the type of atom. This technique could identify only chemical elements, not compounds, so the broader interpretation of the data relied, once again, on well-informed but possibly flawed theoretical models. The presence of hydrogen peroxide, or indeed any of the volatile substances supposedly responsible for the GEX, LR and PR data was not specifically confirmed.

To this day, there remains substantial doubt about the exact constituents of the soils of Mars, whether living or dead.

A TEAM DIVIDED

The PR and LR experiment's Geiger counters could quantify the presence or absence of carbon-14 down to individual atoms. However, as we have noted, they could not identify the kind of molecules in which they were embedded. A molecule of even the simplest amino acid, glycine, $CH_2(NH_2)COOH$, might whizz by and register as no more than two carbon blips. The obvious question is, why couldn't the vapour products of the GEX, PR and LR experiments be pumped to the GC/MS for more detailed analysis?

Partly, this came down to the sheer complexity of Viking's miniaturized hardware, and a dozen practical details to do with the individual procedures. Only so much could be done. Another factor was strictly human. The lab as a whole may have been built all of a piece by TRW, but its various procedures had been invented by a number of different scientific teams. As we have seen, these teams did not always work in perfect harmony. The GC/MS, with its own box and soil hopper, was metaphorically as well as literally compartmentalized from the GEX, LR and PR experiments, which shared an adjacent box.

Viking's biologists had been prepared to swallow a few oddities in the data as long as the GC/MS found some organics in the soil. The Mars mission had failed to deliver its most important piece of evidence. So a non-living, non-organic chemistry was presumed responsible for all the findings.

By the end of 1976 the hydrogen peroxide theory held sway, for all its flaws. Complex models were put forward to explain its formation in the atmosphere from the ultraviolet disruption of water vapour. Alternatively, it may have been produced from the vapour by the catalysing action of metal compounds in Martian dust. If all the water now in the air condensed at once, it might form a layer about half a millimetre deep on the surface of Mars. Such a small quantity could not have been responsible for saturating the ground with peroxides to a depth of, perhaps, several centimetres. So, the present-day peroxides in the soil must be very old, formed gradually over a long period. This means the peroxides, being inherently unstable, should have decomposed by now. The current theory suggests that grains of soil are coated in iron oxides which protect peroxides in their interiors against ultraviolet radiation. The Martian dust might provide such a covering, since it consists of particles so tiny they cannot properly be described as dust. They are known as 'fines'.

In November 1976, Mars slipped behind the sun relative to earth, and the four Viking craft—two orbiters and two landers—were sent into a scheduled electronic slumber for a few months, until a radio line of sight could be re-established. Viking's senior scientist, Gerald Soffen, summed up the situation. 'This entire experience was the greatest joy,' he said. 'How do we pay back the taxpayer for letting us search for life on another world? But in another way it was the greatest tragedy, in that we found no life on Mars.' Nasa's official press releases were carefully phrased, stating that the experiments had 'neither rigorously proved nor rigorously disproved the existence of life.'

One of the experimenters didn't agree.

A question arose about the results from the GC/MS. To be more specific, Gilbert Levin actually challenged the instrument's reliability. Was it capable of detecting very small quantities of organic material, as one might find in, say, the thousand microbes which may or may not have been responsible for the consistently positive LR and the initial PR data?

Levin and a colleague, Patricia Straat, argued that the GC/MS wasn't up to its job. Maybe its failure to detect organic compounds didn't actually *count* as a valid finding? In that case, the LR results were of critical importance, and the hydrogen peroxide theory might be entirely misleading.

While Nasa presented a common front, putting out press releases 'from the Viking Biology Team', a bitter behind-the-scenes row was brewing. Levin departed from the official party line, and claimed that life on Mars had in fact been detected. He began a three-year hunt for proof that the GC/MS sensitivity was inadequate. As it happened, the evidence he was looking for was already tucked away in his own papers, but at first he didn't realize it.

Before the Viking's experiments were loaded on board the spacecraft for their long flight to Mars they had, of course, been tested on earth. A variety of samples was analyzed, from warm sods of Californian lawn to frozen cupsful of Antarctic tundra. Exposed to these benchmark samples, the GEX, LR and PR tests all demonstrated fine sensitivities. The subsequent performance of these instruments on the surface of Mars was never seriously questioned—at least, not until minor hardware failures at a late stage (leaking gas pipes and so forth) produced obvious flaws in the data. The GC/MS was a different matter.

One particular terrestrial soil sample, Antarctic No. 638, was tested in another laboratory, using different instruments, prior to its submission to the GC/MS calibration procedure. This soil was known to contain organic compounds. Klaus Biemann supervised the eventual GC/MS calibration test against 638. In 1979, three years after the Viking landing, Biemann reported, entirely legitimately, that organic compounds had been found in 638, thus helping to confirm the GC/MS's fine level of performance, despite the hiccup with the robot arm at Chryse. In his overview he mentioned that organics had been 'undetectable' in another Antarctic sample, No. 726. This wasn't quite the same as saying there *were* no organics in 726, but Biemann did not quite clarify the issue, and it seemed of little importance at the time. The GC/MS appeared triumphantly vindicated by its analysis of 638. If it said that no organics were present on Mars, then it had to be believed, because the 638 sample had been regarded as a tough test for any instrument.

All the calibration tests undertaken prior to the actual space mission by Viking and non-Viking instruments alike remained on record for anybody who cared to look at them, but first their suspicions had to be aroused. It occurred to Levin that he should check to see if his LR experiment had ever analysed a portion of 726, the Antarctic sample which Biemann seemed to have skipped over so casually in his 1979 statements. So many hundreds of soils from so many locations had been tested in all the Viking gear, including the LR. After a busy four or five years, Levin couldn't quite remember whether or not 726 had been among them.

In fact, it had. As it happened, 726 was a particularly tricky sample. The independent test procedures revealed traces of organic carbon inside microorganisms at levels of less than three hundredths of one per cent of the sample's overall bulk. Atom by atom, the LR also found them when its time came, which is why Levin never had any particular reason to remember 726. The LR was capable of detecting the presence of as few as ten microbes in its samples, and did not fail any of its calibration tests. At least, this is what Levin has claimed, and nobody seems to have contradicted him to date. It was only because Levin was double-checking the calibrations of Biemann's GC/MS that any anomaly cropped up.

Obviously the GC/MS hadn't done so well against 726. That explained Biemann's unwillingness to dwell overlong on this particular sample when backing up the accuracy of his machine. Levin bears no grudge against him today. After all, Biemann had been among the first people to raise the possibility that his GC/MS might not have been able to detect very small numbers of Martian bugs. As for 726, Biemann may have believed that if the GC/MS hadn't found anything, then maybe there really was nothing to find.

The point that Levin makes today is that a deliberately competitive *comparison* of the various Viking instruments had not been conducted with sufficient rigour before the mission. 'Klaus was a decent enough man,' Levin says today, 'but that doesn't change the facts. Before it ever went to Mars, the GC/MS was an impugned instrument. Had the facts [about 726] been published and recognized, the negative GC/MS results would not have weighed so heavily against the positive LR results.'

Cross-comparisons of the GEX, LR, PR and GC/MS before flight could have enabled their subsequent results from Mars to be interpreted in a single theoretical model. The discrepancies in the four sets of data have not properly been resolved to this day because there is still no common mean level of calibration to set them against. Meanwhile, Gilbert Levin asserts today that his LR data from the Viking missions 'constitutes strong evidence for the existence of microorganisms on Mars'.

ULTRAVIOLET DESTROYS AND CREATES

There was another irony buried within the pre-flight calibrations of the experiments. We mentioned that the Pyrolytic Release experiment, the PR, simulated Martian sunlight with a xenon lamp, but with a filter to cut out ultraviolet in order to 'prevent certain peculiarities in the test results'. Horowitz and his colleague Jerry Hubbard initially ran their PR simulations using full-strength sunlight at all the wavelengths to which the Martian surface might be exposed, including ultraviolet. Horowitz, after all, was always keen to stress that the PR was designed to be 'just like Mars'. However, the ultraviolet content of the fake sunlight caused some embarrassing problems in the procedure. It acted on the simulated atmospheric mix of carbon dioxide, carbon monoxide and water vapour, disrupting hydrogen from the water, separating out some oxygen atoms and recombining some carbon to create acetaldehyde and glycolic acid. Both of these are hydrogen-oxygen-carbon compounds usually associated with life.

Another of the products inadvertently generated is comfortingly familiar even to young schoolchildren, if only because they try to avoid it. Formic acid (HCOOH) gives nettles and ants their sting. Horowitz proved that it could just as well be manufactured on Mars as in a back garden on earth. Ultraviolet, which was supposed to be destructive to Martian organisms, turned out to be quite capable of synthesizing some of their ingredients—just as it once did on earth.

Horowitz blocked the ultraviolet from the xenon lamp aboard Viking, so that his experiment wouldn't be capable of manufacturing its own 'astounding' results. It was not lost on him, of course, that his pre-flight test had actually delivered a genuinely astounding result of its own. Before 1976 he was pleased to have found out, well ahead of other scientists, that ultraviolet could manufacture as well as destroy some of the organic ingredients of life out of nothing more than a primitive planetary atmosphere. After 1976 this triumph became something of a nuisance, providing Gilbert Levin with yet more ammunition for his theories. Horowitz remained convinced that Mars was lifeless, even though his own experiments consistently tended to betray him.

Peroxides and ultraviolet: were these two great monsters working in tandem to destroy every prospect of life on Mars? They were not. If either of these assumptions was true, we'd probably not be here to question them. To learn more about modern Mars we must look to the origins of the solar system, the formation of the planets, and the early history of the earth. We must reach back four and a half billion years to a time when Mars and earth, so different today, might well have been as alike as two peas in a pod.

CHAPTER FOUR
THE DAWN OF LIFE

ACCORDING TO OUR best estimates, the sun is somewhere around five billion years old. Its birth followed an extremely long process. First, vast clouds of hydrogen and helium had to condense from the dead remnants of earlier generations of stars. The clouds drifted through interstellar space for countless millennia, tugged about by galactic tidal forces.

Somewhere within this cloud, a seemingly insignificant clump of matter coalesced, gaining mass until it was capable of attracting other nearby atoms and molecules. The gravitational forces involved were so delicate as to be virtually non-existent. But not quite. Over time the little clump of matter grew to the point where its gravity became strong enough to pull in more material from distances of several hundreds of kilometres.

A 'protostar' had formed.

After more millions of years, a large region of the cloud was beginning to spiral inwards towards the embryonic star. The spin forced this region into a disc shape. The protostar's mass increased. So did its gravity field. Its reach extended billions of kilometres into space, pulling in yet more matter, building up greater gravity and piling on the pressure at the protostar's core. Eventually the pressures and temperatures became so

THE DAWN OF LIFE

intense that matter could no longer exist in any conventional state. The core ignited in a gigantic thermonuclear reaction, and our sun exploded into life.

Meanwhile, perhaps a dozen brother and sister suns could have been created in similar fashion from the original interstellar cloud. Our sun and one of its young siblings might have snagged each other's gravity fields, locking themselves into the eerie waltz known as a 'binary' system, with both stars circling a common centre of gravity somewhere in the space between them. There are countless such binaries in the cosmos. As it happens, our sun escaped from the nursery as a loner, finding its own way in the galaxy, and carrying with it the remains of its birthing disc of gas and dust. By now, this disc was starting to condense into special sub-structures of its own.

At least one of these secondary formations tried to become a star in its own right, gaining mass and building its own immense gravitational field—but never so much that it could create the high temperatures and pressures at its core which a true star needs. Had it been able to sweep up hydrogen from the disc for a little longer, then it would have burst into light and challenged the sun for supremacy. Another kind of binary star system might have been created, if only the surrounding supplies of dust and gas hadn't run out just as the competition was getting into its stride.

But binary systems are not healthy places for planets to live in, nor for life to develop. Our single sun is notable for its ability to support its family of nine worlds in neat and stable orbits. A two-star system would have hurled its planets around like hapless ball bearings on a pinball table. None of them could have achieved the long-term stability required for life to gain a foothold. Had the sun's rival ever succeeded in catching fire, we humans would not be here to tell this story and provide, at least, the dignity of a name to the beaten challenger—Jupiter, now the fifth and largest planet in the solar system.

The radiation of the successful new star boiled away the lighter hydrogen molecules in the surrounding disc, so that the regions closest to the star became, by default, richer in heavier elements like iron and silicon, while the outer regions were able to hold on to hydrogen. This division by density was possible only because the original gas cloud contained a variety of substances to start with. The very first generations of stars, created at the dawn of the galaxy, would have consisted of nothing but hydrogen and helium, the simplest units of matter. All the other elements, the basic chemicals which create the richly variegated structures of our world today, had to be manufactured in such suns as by-products of nuclear fusion.

Those suns then had to scatter their products in the vast clouds from which our modern solar system was born.

Some stars would have pulsed out great waves of these chemical products from their outer layers. Others had to die, exploding and showering vast regions of surrounding space with their material. Hydrogen gas remains the most copiously distributed substance in the cosmos, but infinitesimal hints of other elements are mixed into the nebulous interstellar brew.

The grading in the solar system's building materials may explain why some of the emerging substructures—the inner planets, Mercury, Venus, earth and Mars—condensed out of the disc as worlds made principally of silicon-rich crusts, with hearts of molten iron; while the failed star Jupiter, along with Saturn, Uranus and Neptune, emerged as gaseous giants with no rocky crusts, little if any iron content, and interiors made of hydrogen (although these cores are immensely compressed).

We know the least about Pluto, the ninth and outermost world. Its orbit is eccentric, sometimes bringing it closer to the sun than Neptune, the eighth planet. If we were eventually to discover that Pluto has an iron core, like the inner planets, then our models of solar system formation would be upset. It seems more likely that Pluto is not massive, perhaps not even a full-scale planet at all, but a large asteroid, or a moon which escaped from its former orbit around Uranus, the seventh planet.

Mars and the earth, at any rate, fit rather more tidily into the accepted model for planetary formation. Their early lives may well have been very much alike.

The earth's outer crust solidified approximately four and a half billion years ago. Exactly where, when and how life first took a hold upon this new surface, we cannot be sure.

THE DAWN OF LIFE

THE EARLIEST TRACES OF LIFE

The biggest problem for evolutionary scientists who wish to study the dawn of life on earth is that the earliest organisms would have been too delicate to leave their imprints in the fossil record. To complicate matters, the crust itself has been reshaped many times in its history by volcanism, tectonic shifts and erosion. The modern crust, including the places where we dig for fossils, is relatively young, which is why so few traces remain of the countless meteorite impacts which once battered our world just as thoroughly as they did the moon and Mars. Earth's oceans, its ceaseless winds and rains, and the ubiquitous chemical and physical actions of life have applied a deceptive layer of make-up across its surface.

However, there are a few genuinely ancient specimens of crustal rock still surviving today. Just off the south-western coast of Greenland there is a bleak, jagged island called Ikilia. Few tourists would wish to visit here. This freezing cold, windswept place seems thoroughly inhospitable, but for the right kind of visitor it is truly captivating, for it boasts the oldest known geological features on all of the earth.

There is a particular formation which the Greenlanders call 'Itsaq', or 'Ancient Thing'. It was once the soft, muddy floor of an ocean, and over time that floor has been pushed above sea level by the earth's restless interior. The layers of sediment have long since been compressed and hardened into rock. And in these rocks, there may be indications of ancient life; not fossils, to be sure, and not even the merest indication of what that life might have looked like, but rather, the subtle remains of its chemical impact on the surrounding environment.

The Itsaq rocks were laid down nearly four billion years ago, when the earth was barely one ninth of its present age. The signs are that, by then, biological forces may already have taken hold. In the summer of 1996, a team from the Scripps Institution of Oceanography in California, led by Gustaf Arrhenius, analysed carbon compounds in the Itsaq rocks to determine the mixture of carbon-12 and carbon-13 isotopes.

We have mentioned that organic chemistry tends to favour the lighter carbon isotope over the heavier. The difference between one isotope and the other, statistically averaged out across a sample of millions or even billions of atoms in a scrap of organic matter, can be as little as three or four per cent. Yet, within living organisms, this ratio remains very consistent.

Once an organism dies and is fossilized, the isotopes can be dislodged by erosion, chemical decomposition and other forces, and the precision of the original ratio becomes increasingly degraded. Even so, an investigation of supposedly 'fossil' chemical compounds may show a carbon isotope ratio consistent with other samples from around the world which are *known* to be true fossils. Then, scientists are inclined to believe that the compounds must originally have been created by living chemistry. The relevant tests are conducted using mass spectrometers, advanced versions of the instrument which Viking took to Mars.

The Itsaq rocks contain carbon compounds which could be dismissed as natural mineral products. Apatite is a common enough compound of calcium, carbon, oxygen, phosphorous and fluorine found in many rock formations around the world, and particularly in far northern latitudes. Significantly, it is also associated with life. The outer enamel of our teeth, for instance, consists largely of apatite. Hence the dentist's traditional preoccupation with fluoride-enriched toothpastes and fluoridated water. (Fluoride combines with apatite to form fluorapatite, so giving teeth better resistance to bacterial decay.)

According to the Scripps Institution's data, grains of apatite exist within the Itsaq formation. The carbon isotope ratios in these grains seem to suggest that the apatite is an ancient and much decayed by-product of life, perhaps from some organism that made use of available mineral surfaces and exuded calcium compounds during metabolism. The Itsaq fossil apatites, if that's what they are, cannot be considered even as the most degraded versions of the original organisms. Rather, they are the subtle echoes of their ancient wastes.

Until now, apatite in rocks doesn't seem to have been tested in this way. If the Scripps data bears up to scrutiny, then this would place the dates for the origins of life at least 300 million years earlier than anybody has so far suspected. However, push the date much further back, and the earth would scarcely have been solid yet. There is a growing suspicion among the world's scientific community that life didn't have to struggle for domination. It got going almost as soon as the crust had formed.

If we stretch that far back in time to four billion years ago, we have good reason to believe that earth and Mars may then have been quite similar. They were born from the same bloodline of inner planets, as we have seen, and they would have started in life with a shared kit of ingredients: solid silicate crusts, molten iron cores and hydrogen-rich atmospheres, all exposed to powerful ultraviolet radiation from the sun, which triggered a number of essential chemical disruptions and recompositions.

The oxygen which dominates and sustains life on earth today, and which makes our modern planet so different from Mars, is a relatively recent phenomenon, a product of the first great wave of atmospheric pollution. Whatever life forms may or may not have left their traces in the rocks of Itsaq, they would not have breathed the same air as we do today. The primitive air was created largely by volcanoes, belching out compounds based around hydrogen: specifically, methane (hydrogen and carbon, CH_4), ammonia (hydrogen and nitrogen, NH_3) and water vapour (hydrogen and oxygen, H_2O), the latter condensing to form the oceans. The sun's ultraviolet radiation quickly set to work fabricating new ingredients out of this mix, breaking up water molecules to liberate oxygen, which in turn attacked ammonia to liberate nitrogen, or methane to create carbon dioxide. Even more water vapour emerged as a by-product of these processes, but quantities of lone hydrogen atoms were left unpaired. Lighter than any of the surrounding compounds, they drifted up through the atmosphere and were eventually lost to space. So, the atmosphere transmuted almost entirely into carbon dioxide and nitrogen, with water vapour clouds.

The ingredients for life were firmly in place. After all, even something as complicated as a human body is made up principally from just four elements: hydrogen, oxygen, carbon and nitrogen. Together, these account for 96 per cent of our body weight. The remaining 4 per cent is taken up by calcium, with hints of sulphur, sodium and traces of iron, copper and magnesium. All of these were available to primitive biology from the moment that the new oceans first began to wear away at the earth's mineral-rich crust.

All of the processes at work during the early life of earth may have been active on Mars at the same time. Even today, there is still a 3 per cent ratio of nitrogen remaining in the Martian atmosphere, and of course plenty of carbon dioxide. Neither of these gases would have been the original product of volcanism, but ultraviolet-powered fabrications from the atmosphere, formed at a later date. Water vapour is still present, and a variety of minerals exists in the soil.

Such primordial conditions were simulated in miniature by Stanley Miller and Harold Urey in 1952, during one of the most important scientific experiments of the 20th century. Their procedure was very basic, consisting simply of a large jar containing water vapour, ammonia, methane, carbon monoxide, carbon dioxide, nitrogen and hydrogen cyanide, in a replication of earth's earliest environment. Electric sparks simulated lightning in order to trigger chemical metamorphoses. (An ultraviolet lamp would have served just as well.)

After only a few weeks, Miller and Urey's little atmosphere had started to transmute in the expected way, but it also produced a brownish stain on the sides of the jar. This turned out to be a rich broth of organic compounds. Most amazing of all, there were some amino acids, the building blocks of proteins, and thus of life. The brown tar was a long way removed from actual proteins, and as far from DNA in its complexity as a sparkplug is from a motor car. Nevertheless, an important truth had been demonstrated: the geologic conditions of early earth were quite capable of kick-starting organic chemistry. Even so, we must bear in mind that the Miller-Urey experiments were not nearly sufficient to explain how *life* emerged on earth. They simply manufactured some of the essential compounds.

Though we cannot begin to guess exactly what the first organisms might have been like, or quite how they emerged, we can be reasonably sure of the most primitive surviving examples today: blue-green algae. These simple organisms behave in some ways like plants, in that they photosynthesize. However, they are classed as bacteria. With their modern behaviour and internal chemistry as a template, and knowing that the free oxygen in today's atmosphere cannot have been an original feature of the earth's environment, we have been able to piece together some of the first great upheavals in the story of terrestrial life.

Nearly four billion years ago (Itsaq suggests at least 3.8) the oceans supported a great mass of blue-green algae. Pigments in their structures protected them against ultraviolet radiation, but admitted certain wavelengths of light so that the algae could perform photosynthesis. After absorbing carbon dioxide, the blue-greens 'fixed' the carbon into their chemistry using sunlight to power the reactions. Then they excreted the unwanted oxygen, incorporating just a little into their molecules along the way, but only through a cautious chemistry designed to protect them against corrosion by excessive oxidization.

The product of these first biological factories was a form of global atmospheric pollution as destructive as any eco-horror we may fear today. The blue-green algae became a victim of their own success, pumping out vast quantities of oxygen gas, and changing the earth's atmosphere into a lethal mix that they could no longer breathe.

Living on sunshine, carbon dioxide, nitrogen and a few handy minerals, the blue-green algae did not need to eat each other in order to obtain their organic fuels and building blocks. Nor were there any other organisms sufficiently advanced in the uses of energy to prey upon them. The main threat to survival was radiation, not predation. The same ultraviolet from the sun which supposedly sterilizes the soils of Mars also plagued the ancient earth. Hence the protective pigments in the algae.

However, as Norman Horowitz discovered during the preparations for Viking, the ultraviolet probably contributed organics as well as destroying them. Any compounds which the radiation had made would not necessarily have been *un*made straight away, especially if atmospheric products synthesized shortly before nightfall happened to sink down into the oceans before the sun came up again next morning.

Although we should not regard it as automatically lethal, there's no doubt that ultraviolet can be very damaging to life

THE EXPLORATION OF MARS

forms exposed to it. Ultraviolet contributed a wide variety of hazards for primitive life on earth, including a poisonous chemical which it manufactured from the atmosphere, along with all the other compounds it was assembling and destroying.

Enzymes are proteins which aid essential chemical reactions without themselves being altered. When we examine the many enzymes in a biological entity, we can deduce that two of them—the catalaze and disrubinaze enzymes—must originally have served a very different purpose from the one they perform today. Currently, they are involved in oxygen respiration; but billions of years ago, oxygen was expelled, not sucked in. Among other useful properties, catalaze and disrubinaze have the ability to destroy a substance whose formula is written as H_2O_2. Hydrogen peroxide.

PEROXIDES AGAIN

Hydrogen peroxide once constituted one of the greatest threats to life on earth. At a time when there was little free oxygen in the atmosphere, necessarily there was very little ozone, a compound made (again by ultraviolet) out of three oxygen atoms—when they happen to be available. Ozone blocks ultraviolet radiation, but it did not exist in significant quantities three billion years ago. As a consequence of this shortage, water vapour in the air was attacked by the ultraviolet, which slammed into the molecules and smashed hydrogen atoms away. The left-overs consisted of 'free radicals', oxygen atoms each connected to one hydrogen atom when they should have been tied to two. Thus the radicals had unpaired oxygen electrons dangling, as it were, in space. They were extremely keen to connect with something. The obvious course was to latch onto another hydrogen-oxygen free radical. The result was H_2O_2, hydrogen peroxide, which quickly became plentiful in the primitive environment.

This peroxide then attacked the blue-green algae, whose defence consisted of the catalaze and disrubinaze enzymes. The iron-rich catalaze, in particular, broke down the peroxides and shed the oxygen (just as metals on Mars are supposed to have done). In time, as the ozone layer built up and reduced the ultraviolet onslaught, the peroxides on earth dropped to mini-

THE DAWN OF LIFE

mal levels, and the defensive enzymes were eventually adapted for other purposes.

This brings us neatly back to the Viking debates of 1976. While Norman Horowitz and Vance Oyama were arguing that life on Mars could not exist because of the supposed presence of hydrogen peroxide in the soil, Gilbert Levin argued just as vociferously that the peroxides were a myth. To round off his side of the debate, he was able to demonstrate that even if they *did* exist, they could do so only because there was once a great deal more water in the Martian environment for ultraviolet to work on. That would imply plenty of original hydrogen and oxygen ... and so on. Life on earth had found a way of living in a world drenched with peroxide, so why shouldn't Martian bugs have done just the same? Maybe something like the catalaze or disrubinaze enzymes was still at work out there?

Levin completely upended Horowitz's and Oyama's arguments, so that, in his version, the peroxide theory favoured life just as much as it counted against it. Admittedly, there is little proof today that Mars ever managed to create an effective ozone shield. Modern life forms would have to bury themselves under the surface to get away from the ultraviolet.

So far, despite all the ambiguous news from Viking, there is nothing to rule out the emergence of life on the Red Planet. The surprise is not that earth and Mars might once have been quite similar, and maybe both capable of generating primitive life four billion years ago. Rather, the puzzle is why these worlds should have ended up as different as they are today.

WHEN WORLDS DIVERGE

Four billion years ago, when the Greenland rocks of the Itsaq formation were first laid down as seabed silts, earth and Mars might well have been following parallel courses. It is reasonable to suppose that their basic chemical stockpiles were equivalent, but there must have been other factors at work. Particularly temperature. How close can a planet orbit the sun before high temperatures make life impossible? For instance Venus seems very earth-like in some respects, except that it is far too hot. A runaway 'greenhouse' effect within its thick carbon dioxide atmosphere has trapped heat so that the planet's surface today makes Hell seem chilly. More importantly in the case of Mars, how far out can a planet's orbit reach before it becomes too *cold* to support life? Mars' orbit takes it 55 per cent further away from the sun than the earth, but we cannot be sure whether or not this is too far away to have allowed for the emergence of life.

Another problem is that Mars' gravity field has, perhaps, never been strong enough to hold down a dense atmosphere. Much of the gas will have escaped into space, including so much of the lightweight hydrogen that there may not have been enough left over to build water in sufficient quantities. We just do not know for certain how much water exists on Mars today, but it is obviously not abundant. All we can reasonably conclude is that the evidence of great masses of flowing water some time in Mars' past indicates that things could very well have been similar three or four billion years ago to conditions on earth.

Until something went wrong.

Geologically, the divergence may have been due to the Red Planet cooling faster. Remember that the Mariner and Viking orbiters found only the weakest magnetic field, suggesting that the planet's iron core is no longer a molten mass; nor does it generate the same dynamo effect which gives earth its powerful magnetic field. This quiescence also explains why tectonism has not greatly influenced the fourth planet's modern geology, and why its volcanoes seem to have shut down a long time ago. In general, the reservoir of internal heat which has helped power earth's environment has not been so great a factor in the history of Mars, at least not during recent times.

Another factor is the 'precession' of the Martian axis, which wobbles like the axle of a spinning top in cycles of 50,000 years or so, thereby reversing the northern and southern hemisphere's summer and winter cycles. Just how much this may have affected the chances for life on Mars, we cannot be sure. The earth has not been immune from this kind of wobble itself.

THE 'CAMBRIAN EXPLOSION'

An obvious difference between the two worlds today is that one of them, earth, has substantial quantities of free oxygen in its atmosphere where little existed before, and the other hasn't. About 600 million years ago, the earth's swarms of blue-green algae had become so successful at pumping oxygen into the sky that their survival was threatened. By then, ozone had begun to form from the oxygen, and ultraviolet light was becoming less of a threat. Oxygen poisoning became the biggest single problem. Most of the blue-green algae died off, but some managed to switch around their catalaze and dismutaze reactions to make better use of oxygen, instead of shedding it. Eventually a class of organism emerged which actually thrived on oxygen, skilfully perverting the corrosiveness of this element to derive a new source of energy.

It is likely that some of these new entities stumbled across a way of clumping together as multicellular structures. This may have occurred when reproductive cell division accidentally 'failed' and the error actually turned out to be useful; or else, when one entirely separate organism became entangled with another, perhaps through parasitism, and the *ad hoc* pairing provided both with some mutual advantage. There seems little doubt that the specialist divisions of labour between human muscle, kidney, liver, brain and skin cells owe their origins to partnership arrangements between different primordial species.

The boundaries of species identity between them have, of course, become utterly blurred over time. However, consider the case of *Escherichia Coli (E. Coli)*, a type of bacteria which lives in our guts and aids digestion. These are distinct organisms with a totally non-human genetic identity, yet we and they are part of the same broader collective working arrangement.

The other great driver of the so-called 'Cambrian Explosion' of life was predation. It can sometimes be more efficient, in energy terms, to absorb the material of another organism wholesale instead of waiting around for the slower and more gradual processes of photosynthesis to take effect. The entities which developed this nasty new trick gained fresh chemical supplies very quickly. They grew faster, as long as there were weaker organisms around for them to prey upon. As a result, those organisms at the bottom end of the emerging food chain, having survived the drawbacks of their new oxygen lifestyle, now had yet another problem to deal with. They could survive only by becoming hard to consume.

The new predators, in their turn, had to keep ahead if they were not to lose their lunch tickets after a few generations of evolution by the food organisms. The constant race between eater and eaten became precisely analogous to an arms race. It was vastly expensive in resources, extremely rapid, callously wasteful in the numbers of individual lives sacrificed to the greater aim—raw, unthinking genetic survival—and it delivered massive technical advances within a short time. As soon as an organism evolved to deal with one set of problems, predators would come up with new ones. Threats came not one at a time, but in twos and threes. The evolutionary pace became breakneck. In fact the vast majority of necks *were* broken in the process. Only the most well-adapted organisms survived.

The blue-green algae outgrew their passive ways. Along with multicellular colonies to boost the chances of survival, calcium-enriched protective body shells also developed, enabling large cell colonies to hold together. Crystalline photosensitive 'eyes' gave predators an extra advantage, and muscle tissues emerged whose purpose was purely to facilitate movement through the sea in support of predation and escape from predation. Eventually the sea shores, or the muddy edges of freshwater lakes, provided new and hitherto unexploited niches for any organisms capable of surviving out of water. These evolved into grasses and trees, flowers and insects, dinosaurs, birds, reptiles, mammals, apes—and human beings.

Dolphins and whales started out in the sea, and tried life on land for a while. Ultimately, they thought better of it, so to speak, and slipped back into the water. All their past lives are preserved in their modern forms. Inside their beautifully streamlined flippers one will find the readapted bones of ancient feet and toes (while fish have never had fingers...). The human story is suggested just as poignantly during foetal development. For a short while in our mothers' wombs we develop gill-like structures and other distant reminders of our oceanic origins. Those who doubt the capabilities of evolution since the Cambrian Explosion 600 million years ago to create complicated

THE DAWN OF LIFE

creatures like us might like to dwell on the metamorphosis of a single-celled entity into a human being in less than 300 days.

While the blue-green algae may have dominated earth for three billion years, modern animals and plants have evolved with stunning rapidity during the last 15 per cent of the earth's total chronology. Humans have emerged within the very last one tenth of one per cent of the chronology, perhaps three or four million years ago at most. Our modern selves appeared just 200,000 years ago, probably in central Africa. There are still many doubts about the exact sequence of events, but not even the most devout advocate of the religious impulse can deny—except through unforgivable sophistry—that a parity exists between humans and other animals. A cursory glance at the insides of a mouse or a squirrel will reveal a similar arrangement of bones and organs in miniature. A chemical analysis of chimpanzee metabolism will reveal that we and they are all but indistinguishable.

From microbes into man. Was it a miracle? It was, of sorts. Charles Darwin outlined the principles of evolution in his epic work *The Origin of Species By Means of Natural Selection* (1859) and suggested to those daunted by the unconscious randomness of it all that 'there is grandeur in this view of life'. But privately, he was disturbed by the dispassionate savagery he encountered. In July 1856, having worried for many years about the antisocial, anti-religious and apparently inhumane truths that he was about to unleash upon Victorian society, Darwin wrote to his friend Joseph Hooker, 'What a book a devil's chaplain might write about the clumsy, wasteful, cruel and blundering, low and horribly cruel works of Nature!' Darwin was, of course, the devil's most erudite chaplain, and in the course of his work he lost his faith in God.

Even the optimistic Carl Sagan observed, a century later, that 'the secrets of evolution are death and time'.

Quite apart from death and time, there is another essential ingredient of life we have not yet discussed. Its complexity can not easily be explained by the Miller-Urey scenario. It is the ingredient which gives life its principal information content, controls all its growth and internal processes, and drives the

THE EXPLORATION OF MARS

evolution whose outward effects were so fluently charted by Darwin, even though he could not unravel the underlying hereditary mechanism. According to all our understanding to date, the molecule deoxyribonucleic acid, DNA, actually *defines* the difference between life and non-life. It is the most complex chemical entity we have ever encountered in our explorations, yet it is essentially formed from the simple light elements: hydrogen, oxygen, carbon and nitrogen. Until we have properly explained how DNA emerged we cannot really claim to understand the origins of life.

TURNING MOLECULES INTO LIFE

Protein molecules are responsible for manufacturing the tissues which make up an organism's structure. Enzymes, a special class of proteins, facilitate the great variety of metabolic reactions which sustain life. All proteins are built from a standard kit of about 24 different amino acids, assembled in long chains intermingled with traces of sulphur. Any given protein will probably make use of all the aminos, but it will do so in varying sequences, often tens of thousands of aminos in length.

Some of the aminos might easily be manufactured in a Miller-Urey jar, but stringing them together in protein chains is not so simple as producing them separately. The DNA molecule is required to manipulate the processes which clip the aminos together to make the proteins. Only DNA can perform this trick. Proteins cannot, for instance, be created artificially in a factory, no matter how hard our technologists might try. In turn, DNA is itself utterly dependent upon the surrounding activity of special proteins. So, what makes DNA in the first place? The answer is beautiful in its logical absurdity, like something out of *Alice in Wonderland*. DNA controls its own construction with the help of proteins created by DNA.

This cause-and-effect loop had to start somewhere. It is no trivial matter to discover which came first during the evolution of life on earth: proteins or DNA. This is *the* great unanswered question of molecular biology. The probable answer to this mystery will link us back to the Mars missions—not so much to the Vikings, but to Mariner IX, more specifically to the seemingly fruitless six months that the little ship spent with its cameras inactive, waiting for a planet-wide dust storm to subside ... .

In the wake of the Miller-Urey experiments in the early 1950s, the accepted theory for the creation of life assumed that the oceans became a 'primordial soup' of organic compounds, especially amino acids. The Glasgow chemist Graham Cairns-Smith turned his attention to the problem of building the aminos into long chains, which the Miller-Urey model, for all its success, could not account for. He believed that a sloppy water environment, an ocean for instance, was not capable of creating anything more than random aminos. They would drift about and maybe collide with each other, but they couldn't link up. A mediating catalyst was required, something fairly sophisticated; in other words, something that behaved not unlike DNA. No such complicated organic entity could be assumed in the Miller-Urey soup, since it would itself have required entities of similar complexity to build it. So Cairns-Smith reasoned that something *in*organic must have played wet nurse to primitive biology. At a time when James Watson and Francis Crick had just unravelled the double-helix structure of DNA, and molecular biology was capturing world-wide attention, Graham Cairns-Smith studied simple lumps of wet clay.

As far as Cairns-Smith was concerned, the oceans were not nearly so important in the creation of life as the shorelines, the places where water met land, creating intermediate surfaces which were neither totally dry nor totally wet. Simple sands, he found, were not particularly interesting. The individual grains on a beach are large and unsophisticated. Clays, however, are a different matter. Dry, the grains are extremely fine. Wet, they form a sticky, viscous substance which is like an extremely mineral-rich pool of water, except that it doesn't slosh about and become too diffuse to do anything interesting.

Clays are rich in silicon, an element second only to carbon in the variety of compounds it can help to produce. One of the differences between carbon and silicon is their respective liking and dislike of water. Carbon will float about in water, or in the atmosphere, but silicon prefers a more concentrated mineral environment where it can combine in crystalline matrices. Wet clay is a medium perfectly balanced between these extremes of solubility and density.

At the molecular level, clays consist of crystals rich in silicon, with aluminium, magnesium, sometimes iron, and always the binding agents of hydrogen and oxygen—but no carbon. A wide variety of clay crystals do different things in different circumstances. Some crystals grow together as flat sheets, like stacked roof tiles. Water can pass easily between the layers, and the clays created in this manner are very porous. If they dry out, the flat sheets easily flake away in the wind. Other clays have crystals which bind together more firmly. If they dry out, they don't flake away, but form dense hard-wearing masses.

When water flows through clay, electrons are added to, or shredded from, the crystal surfaces, ionizing them. There is an electrical interaction between the clays and other substances in close proximity to them. Clays are catalysts, facilitating certain reactions around them without themselves being changed.

Graham Cairns-Smith's ideas were not given much prominence when he first proposed them in 1965, but his cause was championed twenty years later by the evolutionary geneticist Richard Dawkins. Combining their thoughts, we can speculate how primitive Miller-Urey molecules found a way of linking into more complex entities using clay as an intermediary.

Given the different behaviour of various clay crystals, their behaviour might have been modified by the close proximity of simple organics. Suppose one clay type had a tendency to loosen in water so that its grains flew apart and it became dissolved. Then, by chance, some grains encountered simple organic molecules drifting in the water. The molecules clung to the grains, and the fortunate result was that the grains became stickier than they were before. The next time they were washed up on a shoreline, the clay grains clumped together and were no longer so likely to be washed away again. Over time, the water splashed in, splashed out, and the weaker clays were carried off, while the organic-enriched clays stayed put. A local density of such clay-organics began to accumulate. The random organic 'soup' from the primordial ocean had found a thickening agent and gained a local 'population density'.

Now suppose that the presence of simple organic substances around the clay grains also affected them when they dried out: perhaps a clay-organic came into being which happened to cling together tightly in water *and* flake apart easily when dry. Such clays and their accompanying organics might have been spread from one part of the world to another on the winds, like grains of pollen. As soon as the grains fell onto the shores of other lakes, they would have dominated native clays which lacked their ability to cling together so well in water. The process of gaining a selective advantage would start over again, especially if the local water contained further supplies of organic molecules (Miller-Urey products) to feed the process.

The petrochemical industry stumbled onto this sort of thing years ago, except that their aim has always been to make clays looser, not tougher. As they drill down in their hunt for oil, they pump organic compounds through their hollow probes. These compounds have the effect of making clays less sticky and easier to push through. There's a type of clay with an exotic-sounding name, 'montmorillonite', which loosens up in the presence of certain organic compounds, yet becomes impenetrable in the presence of others. It also changes its characteristics according to how much or how little organic material is pumped into it. Tannins, bitter-tasting brown organic compounds extracted from tea leaves, and familiar for their use in the leather industry, are excellent clay looseners. Clearly, we have profited from the manner in which clay-organic mixtures do things differently from clay on its own.

It is possible that clays might once have aided the linking of long amino chains into proteins, and that clay-organics may have produced the first self-replicating molecules, capable of passing on characteristics from one generation to the next. After all, this is exactly what clay crystals have always done, albeit in a cruder way, passing on tiny molecular flaws in their structure to the layers above. When those layers flake away, carrying their specific flaws with them, they behave almost like a parent generation passing on key characteristics to their offspring—the new layers that build on top of them once they settle in a suitable new location. The flaws, in turn, may have had subtle but significant influences on the clay-organic interface. A chemical entity might have emerged which was capable of passing on the details of its composition to a future 'generation' in an approximation of the genetic process. The inevitable emergence of different crystal flaws might have helped, in turn, to drive a mutation process in the organics. Most of these mutations would have had no discernible effect whatever, or else would have contributed to a weakening of the clay-organic, so that it washed away for ever. But occasionally, a new clay-organic would have been created which was tougher than its predecessors and yet more capable of replication.

Eventually the organic half of the partnership, enhanced in its complexity with the help of the clays, must have developed replication that was increasingly self-sufficient. If the aminos found a way of becoming proteins, the loop of protein-making-proteins could have been set securely in motion. In addition, protein structures would have given the organics some measure of physical integrity, and they would no longer have needed the clays to hold them together. Instead, they could seek their own fortunes.

Exactly how all this happened remains a matter of debate, but the Cairns-Smith/Dawkins theory is more plausible than the idea that proteins and DNA 'emerged' unaided from the basic amino acid compounds of the oceans.

Dawkins has compared DNA to Stonehenge. We might imagine that the tall, heavy blocks of this ancient structure must have been impossible to hoist upright without some kind of powerful machinery—until we recognize that simple, sloping ramparts made out of dried mud helped the builders to shove the blocks into position and tip them upright. The primitive man-powered technology of the day would have sufficed. Those mud ramps were broken up and carted off long ago, leaving only the *illusion* that the mighty pillars of Stonehenge were somehow raised by magic.

The obvious question that remains is, why can we not see clay-organic partnerships at work today? In some ways, we can, as the oil companies know very well. The study of many catalytic reactions involving clay are of great interest today. Just recently, in July 1996, the *Scientific American* magazine reported, briefly, on the work of James Ferris at the Rensselaer Polytechnic Institute in New York State. Ferris could not quite achieve the magic synthesis of complex molecules out of simple ones; even so, he persuaded 'nucleic acid-derived polymers' (polymers are chain molecules made out of repeating elements) to 'condense around clay-like minerals', thus forming much longer chains than anything yet derived from traditional Miller-Urey soups. The article failed to mention how Cairns-Smith saw this coming more than thirty years ago.

However, the main reason why natural organic chemistry no longer depends on clay is that DNA-dominated organisms have become faster and more efficient at sweeping up available organic resources. Clays are too slow on the uptake. Also, they no longer have the same luxury of time to accumulate organics as they must have had four billion years ago, when they were still greatly superior to the near-life they helped to spawn.

Now we come to the closing months of 1971, when Nasa's Mariner IX arrived in orbit around Mars in time to witness the dust storm which destroyed Russia's Mars III lander. The storm was frustrating for Nasa scientists, who were eager to take a look at the Martian surface. Mariner's spectroscopes took readings from the atmosphere to see what the dust might be made of, and the results, interesting enough if a little vague, were safely logged away. Mariner eventually returned stunning pictures of the surface, and the big buzzword from now on was 'rivers'. The dust was a relatively unglamorous distraction.

Five years later, as the Viking landers scrabbled around in the surface soils, peroxides and iron compounds sitting on the ground dominated the scientific debates. Nobody seemed to be paying much attention to the dust *in the atmosphere.*

Once Carl Sagan's gloom had begun to lift in the wake of frustrating results from the Viking biology experiments, he started, tentatively, to suggest that life-like reactions in the soil were almost as interesting as life itself. With two colleagues, Owen Toon and John Pollack, Sagan retrieved the old Mariner IX spectroscope data from the supposedly 'least interesting' phase of the mission when dust was obscuring almost all the

THE EXPLORATION OF MARS

Martian surface. It was the dust itself which Sagan wanted to investigate. The spectroscope data was not simple to interpret, but Sagan and his colleagues eventually decided that the results were consistent with montmorillonites.

Martian clays.

MARS ASIDE

In all, there is more than enough evidence to support a repeat of the Mariner and Viking missions with a new generation of robot probes. Yet, for nearly two decades after Viking, Nasa seemed to lose interest in Mars.

During the late 1960s, JPL's trajectory experts were starting to look further out into space at the gas giants, such as Jupiter and Saturn. They calculated that earth and the outer planets would all pass around the same side of the sun between 1977 and 1978. This happens only once every 176 years, so it was an opportunity not to be missed. Nasa planners wanted to create a mission which would use the gravity fields of Jupiter, Saturn, Uranus and Neptune to alter the course of a robot probe as it swung by, so that each world would divert the ship to the next. The plan was nicknamed the 'Grand Tour'.

With the probe dragged towards a given planet by gravity, it would pick up extra velocity along the way, and then hurtle around the planet's far side. By mapping out the initial approach very carefully, the flight dynamics experts could ensure that the angle of deflection in each case would enable the ship to continue towards the next planet, with little need for rocket engines. This meant that the flight time of a conventional rocket from earth to Neptune could be reduced from thirty years to twelve, perhaps within the operational life of a hardily constructed vehicle.

The budget analysts instructed Nasa that there were funds to explore Jupiter and Saturn only. The planned 'Voyager' probe, costing half as much as Viking, could not be equipped with instruments for any extended mission. The hardware designers swallowed the budget cuts and carefully constructed the twin Voyagers to be as sturdy as possible.

Just in case.

BAD NEWS? *This grainy television scan from the Mariner IV fly-by of Mars in July 1965 shows the Atlantis region of the southern hemisphere. The craters were a disappointment for scientists back on Earth.*

RUSSIAN DISASTER *The Mars III probe, seen here prior to launch. The top section, a conical heat shield and lander, descended into the Martian atmosphere in December 1971, straight into a gigantic dust storm. The lander did not survive.*

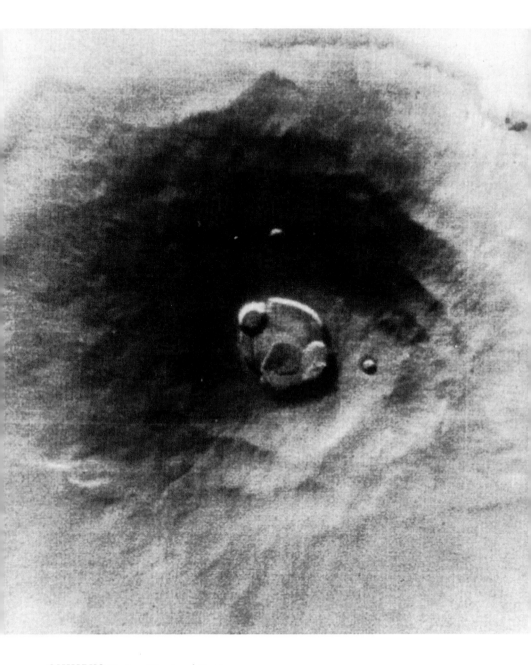

OLYMPUS *Mariner IX entered Martian orbit in November 1971. After waiting for a dust storm to subside, the probe began an extensive photo survey. This view shows Olympus Mons, the largest volcano in the solar system.*

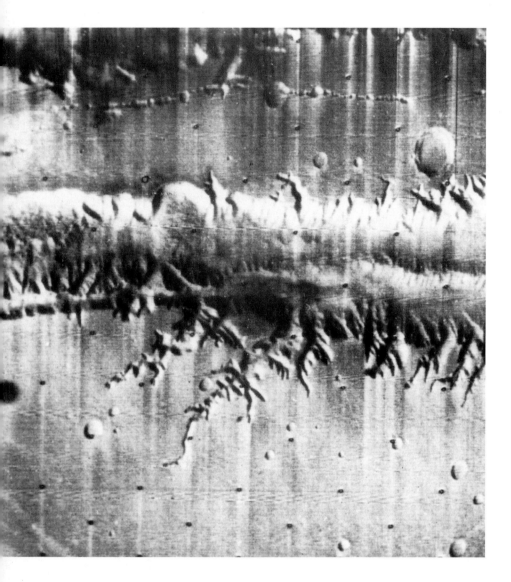

THE BIG DITCH The Vallis Marineris network of canyons was named in honour of the Mariner IX spacecraft which first photographed it in 1972.

BETTER NEWS One of many Mariner IX images of sinuous channels on Mars. This particular example is associated with the Chryse region. The channel's narrow width and the sharp twists and turns in its path seem consistent with water flowing fairly gently across the terrain for a long period. Compare this with the more violent scars of the Mangala Vallis outflow channels as depicted overleaf.

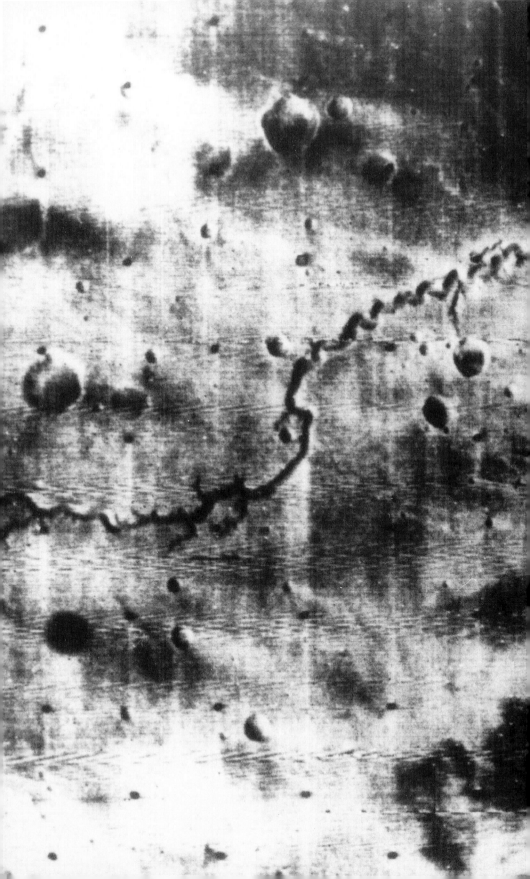

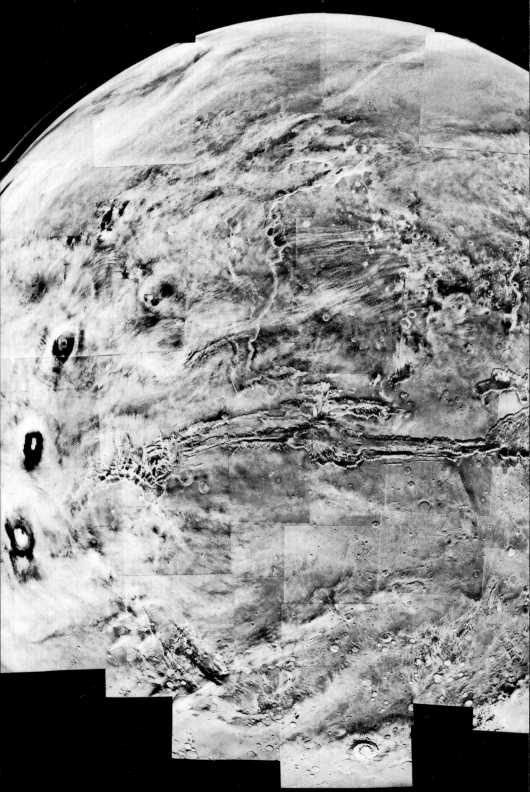

A FASCINATING WORLD *This mosaic of photos taken by Viking I's orbiter shows some of the most dramatic Martian features. At left (to the west) are three of the big Tharsis volcanos, Ascraeus Mons, Pavonis Mons and Arsia Mons. Wispy clouds pervade the extreme northern and western areas of this view. Vallis Marineris, clearly visible here, could span the entire United States from coast to coast.*

WHAT ELSE BUT WATER? *This Viking mosaic of Mangala Vallis shows a complex network of flowing channels, silt deposits and 'islands' gouged out by the flow. The width and scale of these features suggest massive flooding, possibly over a very short period. One theory proposed by Nasa's scientists is that volcanic activity melted and released sub-surface ice.*

ARRIVAL *The first photo from the surface of Mars, scanned just 15 minutes after Viking I's touchdown on July 20, 1976, shows one of the lander's footpads surrounded by sand and jagged rocks.*

THE SMART MACHINE *An artist's impression of a Viking lander on Mars, with its sample scoop extended. When the Vikings actually arrived, they found the surface was much more rock-strewn.*

A WIDER LOOK *Viking I's landing site in Chryse Planitia. Notice the big rock at the extreme left, which was nicknamed 'Big Joe' by the Nasa mission controllers. Dunes of very fine sand are clearly visible, as well as countless rocks. Notice, too, the brightness of the sky.*

INSIDE ALH84001 *A scanning electron microscope image of a tube-like structural form within a carbonate globule from the Martian meteorite ALH84001. It appears segmented, and has an unusual teardrop shape. It is one-hundredth the width of a human hair. It looks just like a fossilized bacterium, but the evidence to support this idea is far from conclusive.*

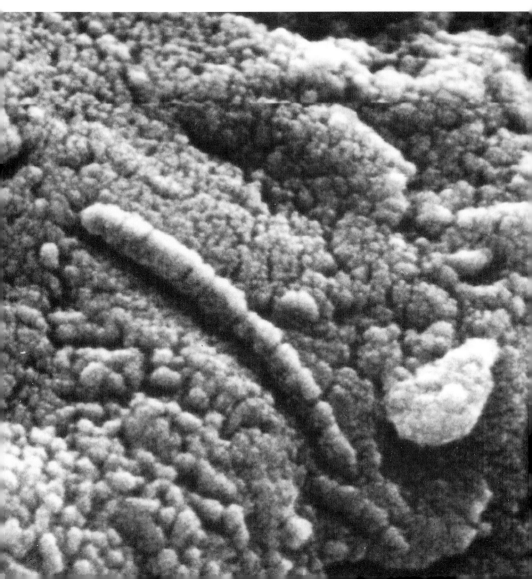

AN EXPENSIVE LOSS *The $2 billion Mars Observer spacecraft failed on August 21, 1993, possibly after an on-board explosion. Conspiracy theorists say that invisible Martian forces were responsible for destroying the Nasa craft.*

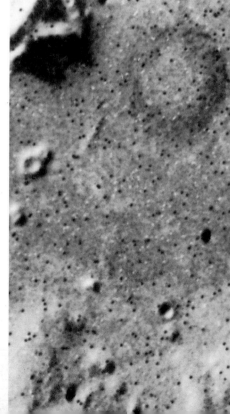

THE FACE OF MARS *The Cydonia region as viewed by the Viking I orbiter in July 1976, showing a rock outcrop which looks very similar to a human face. Highly elaborate speculations about ancient Martian civilizations resulted from this shot. Nasa is committed to rephotographing the 'face' as soon as possible, but does not expect to find anything other than random formations.*

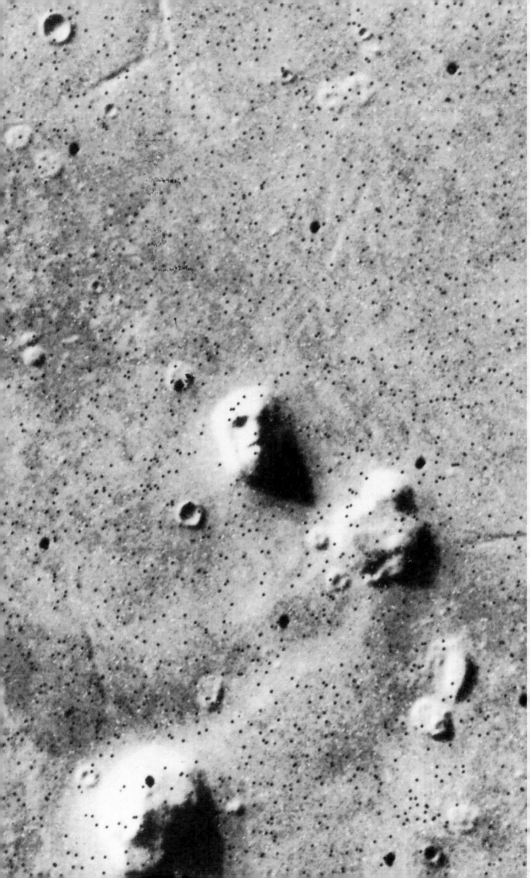

CHEAP INFLATION *An engineer tests a prototype for Pathfinder's inflatable gas-bag landing system.*

STRAPPED FOR SPACE *Pathfinder's solar panels are folded shut around the Sojourner rover prior to launch.*

ON CHRYSE PLANITIA *A scan of Pathfinder's landing site on July 4, 1997. The rover is still stowed on its panel, and a collapsed gas bag can clearly be seen.*

SOJOURNER LOOKS BACK *A remarkable view, taken by the roving Sojourner, of the Pathfinder mothership surrounded by gas-bag material.*

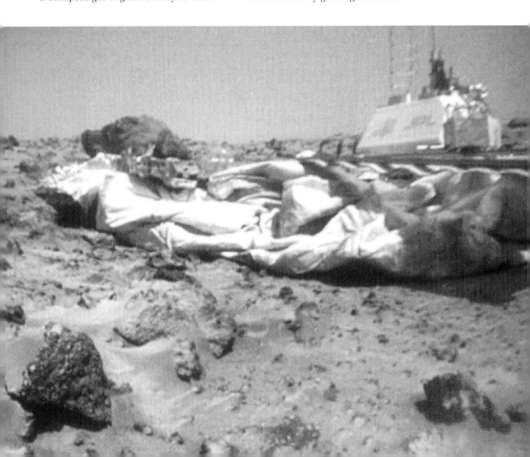

GETTING THERE *This Boeing artwork depicts a 'minimum mission' making use of a small and relatively simple spacecraft carrying a total crew of six astronauts. The upper stage of a landing vehicle can be seen returning to the mothership, while a fresh lander, complete with its living module, descent fuel tanks and protective entry aeroshell, is still connected to the opposite docking port. In theory, such a mission could be mounted in our lifetime for $100 billion .*

The ships were launched a month apart, in August and September 1977. Two years later they swung past Jupiter and returned stunning images of a world in turmoil, with banded clouds, a mass of strange colours, and a permanent red-tinged hurricane that could swallow several earths ....

The original political decision to cut Voyager off at the knees seemed insupportable in the face of these staggering successes. Nasa got its funding and the Voyagers went on to Saturn, where they encountered a ring system more complicated than anyone had expected, as finely grooved as an old long-play record. Some of the finer rings twist around each other like braided ropes as a result of gravitational interactions which we have not yet begun to understand properly. Voyager I was diverted from its greater course so that it could study some of Saturn's moons, while Voyager II continued outwards to Uranus (fly-by in 1986) and Neptune (1989).

After the main planetary encounters were completed, the Voyagers continued to transmit data about the interplanetary environment. In 1996 the decision was taken to shut down the fading radio link with the probes. They had escaped the sun's grip and left the solar system. As Voyager I slipped the bonds of Saturn's gravity and drifted into deepest space, its last picture showed the earth and six other planets as tiny sparks in the darkness, shepherded by a single bright but distant star, the sun.

Since the 'Grand Tour', Nasa has been preoccupied, some say wrong-headedly, with near-earth astronaut activities. Nevertheless, by 1993 a new Mars probe, the Global Explorer, was *en route*, ready to map the planet's surface in very precise detail. On August 21, the probe was getting ready to fire its retro rocket engine in order to slow down and allow the Martian gravity field to capture it.

Then, just as the probe's fuel tanks were being pressurized, Nasa lost the signal. They never heard from the $2 billion Observer again. It seems likely that the ship exploded when its fuel tanks failed.

Dan Goldin, the space agency's new boss, appointed in 1992, told his staff that from now on their big-spending days were over. He demanded that small, cheap ships be designed, costing

THE EXPLORATION OF MARS

no more than a quarter of the Observer's budget. Right now, Nasa's space exploration is engaged on a series of what Goldin calls his 'faster, better, cheaper' projects. Mars is about to become the target of a renewed but cut-price assault. Goldin's new philosophy will be severely tested. Lately, luck has been riding with him. Not the least of his bonuses has been a gift from Mars, delivered right into his backyard. The origins of that gift take us back four and a half billion years to the creation of the planets.

CHAPTER FIVE: METEORITES FROM MARS

EXCEPT FOR PLUTO, whose orbit is so eccentric that, at perihelion, it approaches the sun more closely than Neptune, the solar system's worlds keep to moderately circular paths. These neither intersect nor greatly disrupt each other. However, the solar system has not always been so calm as today, nor its nine planets as safe in their wanderings.

In the 550 million kilometres of space between the orbits of Mars and Jupiter lies the so-called 'asteroid belt', where a vast mass of rubble circles the sun in a diffuse band. Individual asteroids range in size from a few metres to hundreds of kilometres across. None seems quite large enough to have been formed by accretion from the primordial disc, though there is plenty of room for doubt on the matter. It is possible that the asteroids are the debris of primitive planets which did not survive into modern times. One theory suggests that several small young planets condensed out of the disc in the normal way, but developed orbits around the sun that were too elliptical (i.e., very egg-shaped). At some point, their orbital paths intersected. Eventually the planets smashed together, or passed alongside each other so closely that their gravity fields, their tidal forces, tore them apart, so creating a swarm of rubble.

THE EXPLORATION OF MARS

Another possibility is that the asteroids are the remains of a part-formed planet which failed to reach maturity because of gravitational disturbance from Jupiter.

There are many unanswered questions. For instance, we have not yet managed to visit an asteroid in its natural home between Mars and Jupiter. Thus, we haven't yet been able to confirm beyond doubt that most of the recovered meteorites come from that belt. Even now, a small space probe called NEAR—Near Earth Asteroid Rendezvous—is heading for our first formal encounter with Asteroid 433, generally known as 'Eros'. In an echo of Voyager's Grand Tour, NEAR will shoot out beyond the orbit of Mars and then fall back towards the earth, picking up extra velocity during a January 1998 fly-by. The probe is due to catch up with Eros in January 1999.

Eros is about 40km across, irregularly shaped, and just about massive enough to possess its own feeble gravity. NEAR will attempt to orbit it at a distance of only 30km. The manoeuvres will have to be as delicate as a mosquito hovering around the flanks of a lurching bull (though Nasa has plans to smash their little bug all the way into Eros after a year in orbit). Scientists will be able to study NEAR's trajectory and work out, backwards as it were, how much gravity is influencing its flight around Eros, and thus, how massive the asteroid really is. Magnetometers, radar and other instruments will be used to determine whether Eros has an iron core, or is pure rock, or a mixture of both. No doubt some surprises will turn up.

At present, the accepted link between the asteroid belt and meteorites is based on inspired theories rather than absolute proof. We have good orbital models which allow for asteroidal material to reach earth. It seems perfectly plausible that the gravity of Mars or Jupiter might occasionally dislodge asteroids from the main belt and propel them across the solar system.

The mineral contents of meteorites fall into broad categories that one would expect *if* the planetary destruction model were valid. Some meteorites are made almost entirely of iron. This suggests that they originated in an ancient planet whose interior had been iron-rich. Meteoric iron is completely unlike anything found on earth. It has a distinct and highly regular crystalline pattern which can easily be seen when the iron is sliced, polished, and its smooth surface is etched with acid. This so-called *Widmanstätten* pattern is beautiful, but enigmatic. No one has yet explained exactly what kind of crystallization processes could have produced it.

There is also an intermediate class of 'stony irons' which, as the classification implies, is part stone and part metal. Again, they tally with what one would expect from a source planet—a body with an intermediate layer consisting neither of pure iron, nor pure silicate, between the crust and the molten heart.

However, the majority of meteorites are 'stony', very rich in silicate minerals, with only small flecks of metal. Once again, this composition reflects an exterior planetary crust. Many of these stony meteorites are called 'chondrites' because they contain 'chondrules', tiny fragments of magnesium, iron, silicon and oxygen compounds fused into distinctive glassy spheres.

Some theories to account for chondrules suggest that they formed near the young sun, in the surrounding accretion disc of gas and dust, and then became incorporated into planetary and asteroidal matter. Certainly the crusts of the earth and the moon have been so drastically volcanized, smashed into, melted and remelted by geologic activity that any traces of original chondrule material has been obliterated. Chondrite meteorites show us the earliest examples of solid material in the solar system that we have yet encountered.

The survival of chondrules in meteorites is difficult to explain. Any original source body with a rocky crust, an underlying layer of magma and an iron core should have had time to generate some degree of geologic morphism: volcanoes, magma convection currents in the molten interior, crustal recycling and so forth—thus melting the chondrules. But if the chondrite meteorites came from some entity which solidified very fast, locking the chondrules into rock straight away before they could melt, then that object must have shed its heat very rapidly. This implies something much smaller than a planet.

Surely an accreting planetary mass of dust and gas has to be large enough to generate a strong gravity field, pressurize its interior and form a hot magma slurry *out of which* rocky material like meteorites can eventually solidify? The theoretical solution to all these contradictions is known as a 'planetesimal'. This object poses as many questions as it squares away.

CARBONACEOUS CHONDRITES

Among the stony chondrite meteorites, the strangest yet are the 'carbonaceous chondrites', so called because they include carbon in some of their mineral compounds. The carbon is bound up with sulphur, iron, calcium, magnesium and other elements in the form of acid-derived carbonate salts.

These carbonates make the carbonaceous chondrites very intriguing. A carbonate salt will be produced when carbonic acid, H_2CO_3, acts on surrounding metallic minerals. On contact with metal, the hydrogen atoms in a typical acid become dissociated. Metal atoms replace hydrogen in the overall compound to form a 'salt'. As far as is known, carbonic acid cannot occur in nature as a pure substance, although it is commonly found as a weak solution. It's produced when carbon dioxide gas dissolves in water. Rainwater, in fact, is a very weak solution of carbonic acid. The more carbon dioxide we pump into the air during industrial fossil fuel combustion, the more 'acid rain' we create.

The presence of carbonate salts in the chondrite meteorites shows that water once flowed through their cracks and fissures, presumably when the meteorites were still bound up in the greater mass of their original source body. This extraordinary circumstance is just one of the reasons why the carbonaceous chondrites are so highly prized among scientists.

Carbonates similar to those precipitated by acid water can also be produced by the metabolism of certain types of terrestrial bacteria. They live in the fissures of rocks, or in sediments, and are sustained by moisture, creating chemical reactions with calcium, manganese, iron and sulphur. James Fredrickson and Tullis Onstott, working for the US Department of Energy's Subsurface Science Program, have found that mildly acidic water flowing across certain iron-rich minerals will liberate hydrogen gas. This, in turn, provides a chemical power source for the metabolisms of mineral-dependent microbes. It seems quite possible that equivalent ancient bacterial activity left fossil products now contained in vast deposits of manganese ore and reddish-brown iron ore, found in sedimentary rocks around the world. These phenomena are distinct from chalk or limestone deposits, which were created in ancient seas from the calcium-rich skeletons of multicellular organisms, rather than by bacteria in sediments.

For many decades, scientists have presumed that blue-green algae, photosynthetic organisms powered by sunlight, were dominant during the first two to three billion years of life on earth. We are already familiar with modern nitrogen-dependent bacteria living in conjunction with plants, and with the methane belching microbes which live in the guts of animals and among rotting swamp material. However, every month brings new information which suggests the existence of many hitherto unsuspected species of bacteria. Significantly, these species make use of reactions with inorganic mineral materials. It may be that their ancestral equivalents were just as important in shaping our environment as the blue-green algae once were.

Over the last few years, survey teams have encountered new varieties of bacteria in deep-sea locations, around sulphurous thermal springs, and in oil reserves. They have also been found in rock sediments that are so dry and compacted that it seems impossible for living entities to survive there, until one considers that bacteria may have been stranded in sediments which were once wetter and less dense than they are now. Such bacteria can be detected under an electron microscope in a laboratory, but so far all attempts to culture them and make them replicate have failed. They may be just dead husks.

The main reason why mineral-dependent bacteria have been driven into hidden niches is that most of them do not breathe gaseous oxygen. They have not evolved in the same way as the descendants of the blue-green algae, which exploit the oxygen in the earth's present atmosphere, along with the rich organic food supplies within the surface ecology.

It seems unlikely that alien mineral-dependent bacteria deposited the carbonates in the carbonaceous chondrule meteorites. There is no evidence to suggest that this might have been the case. Carbonate salts are not classed as organic compounds in their own right. They are just as easily associated with non-living chemistry. From now on, therefore, the word 'carbonates', will be used in the context of inorganic mineral formations which *may or may not* have been produced by bacteria.

ORGANIC 'PAH' COMPOUNDS

Many carbonaceous chondrites include carbon compounds which are definitely classed as organic. Dark, tar-like patches of material contain Polycyclic Aromatic Hydrocarbons (PAHs). This technical term describes a broad class of compounds which exist in long repeating (polycyclic) chains of benzine-like (aromatic) hexagonal arrays comprised of molecules made from carbon and hydrogen (hydrocarbons). PAHs are often found in close conjunction with the mineral carbonates inside carbonaceous chondrites. This doesn't necessarily mean that the carbonates and the PAHs were formed at the same time. It's just that the tarry PAHs happen to favour coalescing around the carbonates.

For sure, the presence of organics in meteorites tends to generate some excitement among the general public, because the popular press encourages an imaginative leap from organic chemistry to 'life' rather too hastily. In fact the incidence of PAHs in carbonaceous chondrites is so familiar to scientists that they almost take them for granted. Many scientists believe that there is every reason to presume that organic compounds are created in space by the same kinds of ultraviolet interactions which typically generate the classic Miller-Urey products. The solar system's original cloud of dust and gas certainly contained the right ingredients.

Astronomical observations of the vast dust clouds drifting within our galaxy (the Milky Way) confirm that organic compounds are present throughout interstellar space. William Schopf, a biologist at the University of California, says, 'there is nothing spectacular about PAHs. The evidence just tells us how banal organic chemistry is throughout the cosmos.'

Nevertheless, PAHs are often associated with life chemistry whenever they occur on earth. They are what we expect to find after biological entities have decayed. They can also be found in crude oil, or the fumes from car engines.

METEORITES FROM MARS

It might be argued that a car's exhaust products are associated with ancient life, because the engine runs on organically derived fossil fuel. The oily base for this fuel dates back to more than 300 million years ago, to the 'Carboniferous Period', a span of some 70 million years during which the organic detritus from widespread forests and swamplands was compacted, heated and pressurized by geothermal morphism. Yet the PAHs which emerge from a car's tailpipe are far removed from the original biological source. No certain link with life can be presumed *unless* all the other links in the argument are firmly in place. There has to be proof by precedent that oil is derived from ancient organisms. There must also be reliable knowledge of the industrial techniques used to convert oil into petrol, and of the sequence of combustion in a car's engine.

PAHs found within extraterrestrial material have no such traceable history. They are wide open to misinterpretation. Just as PAHs on earth have been greatly altered from their original source, the same is probably true of PAHs in meteorites. The accretion processes which formed the material of meteorites could have trapped organic material from the solar system's dust cloud, then converted them into PAHs. However, the initial *source* is not likely to have been anything biological. So far, PAHs in asteroidal meteorites have not provided definitive isotopic clues to suggest that their carbon may have derived from life.

However, we cannot rule out the possibility that meteorites may have contributed organic material to the earth. There are about 2500 medium to large meteorites in various geological collections around the world, yet they represent only an insignificant proportion of the material from space that has actually hit us. The earth is constantly bombarded by 'micro-meteorites', fragments so tiny that we can seldom recover them. They are small (micro) and lightweight. Friction quickly slows them down when they slam into the earth's upper atmosphere.

As much as a hundred tons of these meteoric dust fragments land on our planet every *day*. As they come to rest among the endless rubble and debris on the surface, lose themselves among desert sands, or sink into the vast oceans, we seldom notice these swarms of alien arrivals, and we cannot yet judge to what extent any organic material lodged inside them, or within their larger meteoritic counterparts, may or may not have contributed to the history of life on earth.

For instance, the Murchison meteorite, a carbonaceous chondrite, fell to earth in Australia in 1969. It was recovered for scientific analysis. Hundreds of organic compounds have been identified within it. These include a quantity of PAHs. There are also a number of compounds more closely related to some of the substances actually used in the metabolisms of living entities. In fact, the most intriguing of the Murchison compounds aren't merely *similar* to fragments of life chemistry; they are utterly indistinguishable. They are aminos, purines and pyrimidines, some of the building blocks of proteins, and even of DNA.

Obviously the simplest pieces of life's chemistry are adrift in the cosmos. The problem is putting them together.

The Miller-Urey synthesis of aminos and other basic organic compounds, powered by ultraviolet radiation from the stars, takes place not only on earth but also on the grandest scale, among the dust and ice clouds of interstellar space. In addition to this intriguing situation, it seems that many meteorites, and the carbonaceous chondrites in particular, must have come from an ancient parent body (a planetesimal or whatever) that was once capable of supporting liquid water.

It's not impossible that some of the inorganic carbonates in carbonaceous chondrites were produced by bacteria, and that the PAHs are derived from their decayed bodies—it's just extremely unlikely. The problem is that the carbonates are not physically quite the same as those produced by bacteria on earth. Anyway, accounting for the acidic water that probably *did* form the carbonates has proved a difficult enterprise in itself. Explaining how biology might have evolved on a planetesimal would be beyond the limits of current scientific knowledge.

If we could only find a sample of rock with carbonates and PAHs that had come from a full-sized planet, one with a dense atmosphere and an acknowledged quantity of liquid water, then the bacterial argument might seem slightly more realistic.

In July 1996, scientists from Nasa reported their discovery of just such a specimen.

SHERGOTTY-NAKHLA-CHASSIGNY METEORITES

The acronym 'SNC' honours the names of three particularly interesting meteorite impact sites: Shergotty, in India; Nakhla, in Egypt; and Chassigny, in France. A dozen SNC meteorites are logged in collections around the world, including six from the Antarctic which have been grouped for convenience into the same class.

The SNCs are significant because they don't fit *exactly* into the stony, the iron or the stony-iron categories. They do not contain chondrules, and they show several signs of geothermal morphism. SNCs are *igneous*—the products of solidified magma. They must have been created on a full-sized planet, rather than a 'planetesimal'.

A wide variety of techniques can be used to pinpoint the time at which the SNC material solidified. Rocks of all kinds tend to contain tiny traces of radioactive elements dating from the origins of the solar system. The metallic element rubidium is a useful example. The rubidium-87 isotope has a 'half-life' of five billion years. Half a given population of rubidium atoms will decay over that period, yielding a different metal, strontium. The ratios of rubidium to strontium in a rock will betray the time that has elapsed since the last solidification, give or take a few hundred million years. According to the rubidium-strontium data, the SNCs appear to be younger than the supposedly asteroidal meteorites. This increases the likelihood that they were created on a full-scale planet.

There is a fascinating complication, though. The dozen SNCs are not all of the same age. Does this mean that they come from different sources? Not necessarily. If the SNCs' source world was geothermally active (as we suspect) then its crust would have produced rocks of varying ages. Our earth provides a fine example of this. The Itsaq formation in Greenland is close to four billion years old, but a number of volcanoes around the world are producing fresh igneous crust at this very moment.

The next step is to uncover a mechanism which can explain how samples from different sites on a source-world could have been propelled into space. Volcanic eruptions might hurl rocks off a planet, given sufficient power. However, it's more likely

THE EXPLORATION OF MARS

that the necessary kind of punch was provided by very large meteorite strikes.

The hunt for the SNCs' home world proceeds by a process of elimination. The solar system has a gravity gradient which tends to favour objects that fall sunwards. A rock fragment smashed away from Mercury would have to work against, and elude, the sun's gravity in order to reach earth. The thick atmosphere of Venus would impede a hurtling fragment on its way up, so that it wouldn't even escape the planet.

Among the outer planets, Jupiter and Saturn are not among the suspects: they are gas giants, not worlds made of solid rock. Jupiter's 16 moons and Saturn's 18 are possible candidates. However, it seems unlikely that fragments from any of these sources could have escaped the immense gravity fields of their parent planets.

Could the SNCs have fallen to earth from *our* moon? Certainly our nearest companion in space has been severely pummelled by meteorite collisions, and it has been volcanic in the past. The moon is rich in igneous material. We do indeed find some meteorites on earth that appear to be lunar in origin, but the SNCs themselves do not match any samples of rock retrieved by the Apollo astronauts in the 1960s and early 1970s during the lunar landing missions.

It has long been suspected that all the SNCs come from Mars. They would have needed only to escape their mother world's gravity field and fall sunwards in order to cross the earth's orbit at some point. The energy requirements are not remarkable. Even so, a mathematical model of trajectories which favour Mars over some other planet hardly constitutes a proof.

In 1979 Donald Bogard at Nasa's Johnson Space Center in Texas studied an SNC meteorite recovered from Antarctica that year. He determined that pockets of carbon dioxide gas trapped inside it possessed carbon and oxygen isotope profiles exactly similar to that of the Martian atmosphere. How could he be so sure? Viking had made atmospheric measurements on Mars three years earlier. The suspect meteorite, EETA79001, had already been classified as a probable SNC. Now there was solid evidence of its Martian origin.

It seems more than likely that SNCs were chipped off the surface of Mars by meteorite impacts. Massive pressure and heat, applied within fractions of a second, produced a variety of 'shock' phenomena within the material. The glassy bubbles which contain the trapped gas were created and sealed up almost instantaneously. Had this not been so, the gas would have seeped out. The SNCs had enough impetus to punch straight through the Martian atmosphere and into space. During their passage upwards, the air rushing by would have cooled the rocky material fast enough to preserve the bubbles.

The shock of meteorite impact has altered some of the SNC material, but by no means all of it. Many undamaged features remain, yielding valuable clues about the original rock formations from which the SNCs were chipped.

These conclusions are more or less agreed among the world's scientists. The SNCs come from Mars. Eleven of them represent samples of relatively recent planetary crust. The youngest appears to be approximately 180 million years old. The oldest first solidified 1.3 billion years ago.

The twelfth SNC sample, only recently identified, dates from a time when Mars was still young.

METEORITE ALH84001

The videotape scenes look like holiday mementoes. Certainly the tape is a good few years old. In fact it was shot on December 27, 1984. An excitable group of people is driving across an Antarctic ice field in a little powered sled. They are obviously having fun. The sled stops and a parka-clad young woman gets out of the sled. The camera's viewpoint jerks downwards to focus on a small dark lump, the size of a potato, lying half-buried in the ice.

'Hey, this looks like a good one!' the young woman shouts.

More than a decade after these cheerful keepsake shots were taken, Roberta Score, a member of a National Science Foundation meteorite hunting team, found herself in front of dozens of cameras in Washington DC, explaining what she had discovered in the remote Allen Hills ice fields of Antarctica. Why had she stopped the sled and instantly noticed that rock?

'The colours looked different. The rock looked very green. It stood out in my mind that it was kind of weird.'

Just how weird was not immediately apparent to the survey team. Meteorite ALH84001 was logged as a probable asteroidal sample. It was sealed into a special canister and shipped to the Johnson Space Center (JSC) in Houston, Texas. For nine years it sat in a protective sealed storage facility, surrounded by nitrogen gas, on a shelf alongside many other asteroidal samples.

In 1993, geologist David Mittlefehldt requested samples of ALH84001 together with several other meteorites from the JSC collection. He was researching possible links with a very large asteroid called 4 Vesta. Slicing into his sample of ALH84001, he found that it wasn't a good match for the suspected Vesta fragments. In fact Mittlefehldt doubted that this meteorite came from anywhere within the asteroid belt. He thought it was more likely to be an SNC.

He saw some interesting orange patches in the rock.

By February 1994, word had spread that an unusual SNC had been identified, and that it was almost certainly much older than the others. David McKay, a scientist at the Johnson Center, had a particular interest in the ancient Martian environment. He put in a proposal to study ALH84001, and his suggestion was accepted. McKay and his colleague from JSC, Everett K. Gibson, were allocated a 2-gram sample.

For six months, an examination at conventional microscopic scales revealed nothing of remarkable interest. Realizing that a more intensive study was required, the JSC scientists joined forces with Kathie Thomas-Keprta from the Lockheed Martin company. The new team switched to more powerful instruments with greater magnification. These had become available only a short while before.

By the summer of 1995, it was clear that the Nasa scientists were growing very secretive about their work on the meteorite. Naturally, this gave rise to rumours at the Johnson Space Center that they had discovered something of exceptional interest. Those who were not in on the secret had to wait a year before they found out what it was.

Soon, everybody in the world would hear about ALH84001.

THE STORY BREAKS

On Monday, August 5, 1996, the journal *Space News* published its usual reports on Nasa's activities, rocket launch schedules and hardware. A snippet item by Leonard David (uncredited), reported on a rumour circulating among the space community. 'Meteorite Find Incites Speculation on Mars Life.'

By Monday night the story had been picked up by the national mass media. The next day, Nasa's chief administrator Daniel Goldin was forced, ahead of schedule, to put out a formal press release. 'Nasa has made a startling discovery that points to the possibility that a primitive form of microscopic life may have existed on Mars more than three billion years ago. The research is based on a sophisticated examination of an ancient meteorite that landed on earth 13,000 years ago.' Goldin was at pains to point out that all the findings were tentative, and that more work needed to be done. 'The evidence is exciting, even compelling, but not conclusive,' he warned.

At the time, David McKay was on a camping vacation with his family, completely unprepared when the story broke early. Once alerted, though, he caught the first available flight to Washington.

On Wednesday, August 8, President Bill Clinton, then on the campaign trail seeking re-election, was preparing for a visit to California. Nasa's sensational news was a real boon to him. Nothing focuses more attention on a president bidding for a second term than appearing before the television cameras to announce a great American achievement.

'If this discovery is confirmed,' Clinton told viewers, 'it will surely be one of the most stunning insights into our world that science has ever uncovered.'

The same day, McKay and his colleagues presented their findings to the press in Washington.

The authoritative technical magazine *Science* was persuaded to release advance prints of one of its articles, in fact the formal scientific paper from the Nasa investigating team. 'Search for Past Life on Mars: Possible Relict Biogenic Activity in Martian Meteorite ALH84001', wasn't supposed to have been published until August 16.

At the press conference, pictures of electron microscope scans were distributed, showing rounded gobbets reminiscent of microorganisms. Several of the shots showed segmented worm-like structures, one hundredth the width of a human hair, apparently frozen into place as mineralized fossils. One night, shortly after this image had first been scanned in the laboratory, McKay's colleague Everett Gibson happened to take a print home with him. 'When I put it on the kitchen table, my wife, who is a biologist, asked, "What are these bacteria?"'

Really the question should have been, *are they* bacteria? This is where many scientists around the world came to disagree with the Nasa team's findings. The late Carl Sagan had a favourite saying: 'Extraordinary claims require extraordinary proof.' This quote was widely used by the critics.

There was, in fact, only one part of the story which every one was agreed upon, if only in principle—the story of how ALH84001 came to earth in the first place.

THE METEORITE'S ODYSSEY

The geologists who studied the igneous rock within the meteorite discovered that it demonstrates all the characteristics of material which solidified at an early stage of Mars' history. The rubidium-strontium dating of the material also supports this idea. The meteorite's material appears to have solidified on the Martian surface more than four billion years ago.

Approximately 15 million years ago, ALH84001 was thrown into space by the impact of yet another, presumably asteroidal meteorite, on the surface of Mars. The evidence for this is twofold. Firstly, ALH84001 bears the traces of a rapid and powerful impact, in the form of narrow fractures and other shock features associated with SNCs in general. Secondly, the outer zones of the rock show signs of exposure to cosmic rays, the powerful, ever-present radiation which pervades all space, possibly as the energetic remnants of supernovae explosions. The impact of these rays on the rock has created a different kind of radioactive clock for geologists to examine.

The meteorite escaped the gravitational pull of Mars and orbited the sun in a manner not unlike an asteroid. However, most of the asteroids have been shepherded into their main belt by the gravity field of Jupiter. ALH84001 was not subject to this kind of control. It probably started its journey in space on a more random trajectory, falling under the gravitational interference of other bodies in the solar system. We cannot be certain how often it may or may not have intersected the earth's orbit. Eventually, however, it passed close enough to be captured by the earth's gravity.

The meteorite fell through the atmosphere and landed in the Antarctic. Heated by its entry, it melted the ice, burying itself deep below the surface. After a short time, the meltwater around it froze once again, sealing the meteorite in place. This way, it was preserved against surface weathering.

We can be reasonably sure that ALH84001 remained locked in the ice for about 13,000 years. The evidence to support this comes from the fact that ice sheet structure is laminated, rather like sedimentary rock. Certain ancient geological events, such as major volcanic eruptions, have given particular layers their own datable characteristics, adding minute traces of chemicals and dust. Scientists have noted, with sadness, how the recent arrival of global industrial pollution is clearly indicated by a marked dirtying of ice layers younger than 300 years.

Antarctic environmental experts understand very well how the ice behaves over time. Old, deep layers are sometimes exposed by wind erosion. In the summer the surface melts, and there are complex movements of the ice sheet as a whole.

There is no particular reason why the Antarctic should attract more meteorite falls than other parts of the earth. The main reason why survey teams focus on this part of the world is that small isolated rocks tend to stand out against the dazzling white ice panorama. When a dark stone like ALH84001 appears in otherwise pristine sheets of ice, far away from any obvious source, it's a safe bet that it must have dropped from the sky.

Around the world there must be countless small meteorites littering the place, disguised to all but the most expert eye as ordinary pieces of terrestrial rubble. They are extremely difficult to identify unless they have gouged out clearly identifiable impact craters. Meteorites exposed to the elements will quickly become weathered, so that they resemble their surroundings. The Antarctic ice fields, relatively undisturbed and clean, have the benefit of isolating and then sealing up meteorites within hours of their landing. Just how good that seal has been in the case of ALH84001 is a matter for debate.

The meteorite has been found to contain PAHs. David McKay and his colleagues claim that these were produced on Mars. They also claim that several distinct orange carbonate patches inside the meteorite are more complex than one would expect from simple carbonic acid reactions. In the prologue to their paper for *Science* they conclude: 'The PAHs, the carbonates and their associated secondary mineral phases and textures could be the remains of a past Martian biology.'

This is where opinions become sharply divided.

INSIDE THE ROCK

McKay's team examined the curious orange patches, which are clearly visible under an ordinary optical light microscope, and about the size of a full stop to the naked eye. They occur on the walls of fractures inside the meteorite. The patches are suffused throughout with the kind of material already encountered in carbonaceous chondrules, which can be deposited by carbonic acid in water flowing across mineral surfaces.

At first glance the carbonate features, which McKay describes as 'globules', *look* more like the sort of deposit often left on rock surfaces by terrestrial anaerobic bacteria (in ornamental garden ponds, for instance). They have an interesting layered internal structure. The inner cores tend to be rich in calcium and manganese; then there is a band of iron carbonate; and the outer layers consist of iron sulphides. This banding of different materials within the carbonate structures could suggest a more complex process of deposition than acidic water activity could achieve on its own.

In addition, tiny grains of magnetite, an iron oxide, are found inside the carbonate globules. Specks of magnetite can be created in the bodies of bacteria which live around rocks, or in silts and sediments. The mineral appears to have some benefit, allowing bacteria to align themselves advantageously in relation to their environment. The grains of magnetite within ALH84001 have surfaces that are suggestive of biology. Everett Gibson argues that the lack of sharp crystal faces on the surface are indicative of a bacterial source.

Mars has only a very weak magnetic field. On the face of it, there is no obvious reason why any native organisms should have evolved little magnetic 'compasses'. However, billions of years ago, the planet's interior had not yet solidified. The magnetic field was probably a great deal stronger in the distant past than it is today.

This interpretation of the grains is extremely contentious. As with all the individual features found within the meteorite, the grains *could* also have been created by non-living processes. In December 1996, a team led by John Bradley at the Georgia Institute of Technology investigating their own little shard of the meteorite managed to slice through one of the magnetite grains, using an intense beam of ionized particles. The team found a kind of internal 'corkscrew' effect in the structure which, they say, is not normally found in bacterial magnetite, although Bradley acknowledges that scientists have not yet had a chance to make comparative tests with a sufficiently wide sample of biologically produced magnetite.

The most controversial piece of evidence was not included in the initial paper for *Science*. It seems altogether too fantastic. At the extreme magnifications delivered by their scanning electron microscope, McKay's team have picked out physical structures resembling fossilized bacteria. They are teardrop-shaped, which is unusual in itself, as if some deliberate gradient of structure were present. Some of the examples appear to be segmented. The trouble is, these 'fossils' are smaller than any bacteria yet found on earth. Carl Woese at the University of Illinois has suggested a further problem. 'These putative Martian bacteria are scarcely larger than the magnetite grains they're supposed to have produced.' He says that the bug-like 'fossils' in the Nasa team's electron microscope are so small, they are more likely to be accidental products of mineral deposition processes than the remnants of living organisms. 'At that tiny scale, it's easy to read significance into random formations', Woese asserts.

But who's to say how small a living entity can be? Thomas Kieft, at the New Mexico Institute of Mining and Technology, has identified what he calls 'dwarf bacteria'. These are starved, semi-dormant organisms in sediments which have become too compacted to support a healthy population. There is no reason why Martian bugs inside ALH84001 shouldn't have produced magnetite grains, and then become shrivelled, and dormant, as circumstances changed, perhaps when the Martian environment altered for the worse and the water disappeared. The supposed fossils might represent dried-up versions of once-healthy microbes. The iron-rich magnetite grains would not have shrunk as drastically as the creatures which manufactured them. Hence, the *apparent* similarity in size today between the magnetite and the putative bugs themselves.

In the same month that McKay and his colleagues announced their findings from the meteorite investigation, Todd Stevens of the Pacific Northwest Laboratory in Washington State announced the discovery of extremely small bacteria living in rock fractures deep beneath the Columbia River. These appeared to be dependent on mineral reactions for their energy supplies. As it turns out, these bacteria are just twice the size of the supposed fossils in ALH84001.

In theory, the carbonates, the magnetite grains and even the strange teardrop-shaped 'fossil' features could have been produced entirely by inorganic chemical reactions driven by acidic water flowing through the cracks in ALH84001 while it was still a part of Mars. However, in their paper for *Science*, McKay and his colleagues argued that varying levels of acidity and alkalinity ('pH' levels) needed to prevail simultaneously in order to produce some of the differentiated materials that they found. A hallmark of biological metabolism is its ability to sustain a range of pH levels within the same system.

In this argument, the Nasa team asserts that the carbonates were indeed created on Mars at a time when water was present. In addition, though, the deposits were facilitated by living bugs.

One of the strongest pieces of physical evidence is that minor fractures run through some of the carbonate globules. Either the fractures were produced when the rock was smashed off the surface of Mars, or they were created on impact with the earth. The point is that some of the carbonate deposits *share* fault lines with the surrounding igneous rock. Therefore the carbonates must have been created *before* ALH84001 hit the Antarctic. It can be assumed that nothing too dramatic would have disturbed the meteorite after it had fallen to earth.

It seems unlikely, however, that the carbonates were created in space. Even in the case of the carbonaceous chondrites, an original planetesimal capable of supporting water (in small quantities, at least) probably produced their carbonates. The obvious conclusion is that the carbonates in ALH84001 may have accumulated while the rock was still a part of a 'wet' Martian surface.

McKay reasons that because the carbonate globules must have been created in the far distant past, when Mars was warm and had liquid water, they are likely to be extremely ancient. He suggests that the carbonates are at least 3.6 billion years old.

However, a rival team of scientists at the Field Museum in Chicago, led by Meenakshi Wadhwa, disagrees. Working on a different fragment of the meteorite, they dated their samples by the rubidium-strontium procedure. Their conclusion was that the overlaying carbonates may be only about 1.5 billion years old. This dates them to a time when Mars was, perhaps, no longer so hospitable to life.

Dating the carbonates is difficult, but important. The 'fossil' shapes which McKay's team think they have identified appear to have been created as part-and-parcel of the carbonate structures and rosettes. The more recently the carbonates and 'fossils' were laid down, the less likely it is that they are associated with life. Just as long as the prevailing attitudes about modern Mars' *lack* of life are correct, of course.

MARTIAN PAHs?

In the summer of 1995, the Nasa investigating team sent a tiny sample of ALH84001 to Richard N. Zare and Simon J. Clemett at Stanford University in California, asking them to look for organic compounds. By then, fragments of ALH84001 were becoming valuable objects. The Stanford team's sample was extremely small. Remember how Viking's pyrolysis chambers heated and vaporized their samples? The tiny flecks of Martian meteorite could not be sacrificed in this way. Zare and Clemett used an instrument capable of dislodging just a few molecules from the main sample. So little material was affected that the bulk of the sample was left more or less intact.

The 'Two-Step Laser Mass Spectrometer' works like this. First a tiny probe is positioned close to the sample. The probe emits an infrared laser pulse of intense power. The beam heats a microscopic area of the sample to 100 million degrees Centigrade, but for only *one ten millionth* of a second. This briefest moment of superheating dislodges molecules from the surface of the sample.

However, these tiny quantities of vaporized products cannot simply be pumped away for analysis. There are just a handful of molecules propelled into the chamber. They must be ionized the instant they become detached from the main sample, otherwise they will simply be lost. Ionization enables them to be carried away by electromagnetic fields and steered towards a mass spectrometer.

To achieve this capture a second laser pulse, this time at an ultraviolet wavelength, is aimed at a point in space fractions of a millimetre above the sample. The ultraviolet beam fires within billionths of a second of the infrared. Thus, a few molecules of the sample are dislodged and ionized almost at the same moment. Different intensities and combinations of laser bursts will dislodge different classes of molecule. (In essence this is an extremely sophisticated version of the stepped heating process that we have previously encountered aboard Viking.)

Using this system, the Stanford team detected PAHs. Zare and Clemett assumed, naturally enough, that their sample must have come from a carbonaceous chondrite. McKay and Everett at the Johnson Space Center deliberately kept vague the origins of the sample *until* the PAH examinations were complete. Zare and Clemett were very surprised to find out that they'd been dealing with a piece of Mars rock.

They were sworn to absolute secrecy until the publication of the formal *Science* paper.

So, where had the PAHs come from? McKay and Everett say that they are the decayed products of the very same bacteria which supposedly produced the meteorite's carbonate globules.

Certainly the PAH profile within the Mars meteorite differs from the patterns associated with carbonaceous chondrites. The ALH84001 compounds are much less varied in their molecular structure, and generally simpler. Bernd Simoneit, a chemist at Oregon State University, compares the paucity of ALH84001's compounds with the comparative wealth found within the most notable carbonaceous chondrites. 'Look at the Murchison Meteorite. It contains hundreds of organics, including the sort of things that organisms actually use, but nobody is saying that there's life in the asteroid belt.'

The poor variation in the PAHs from the Martian meteorite seems like bad news, but it needn't be. McKay and Everett argue that if a population of ancient bacteria did decay inside the rock then the class of PAHs derived from one bug's remains would have been much the same as another's. By contrast, the random variations in PAHs from carbonaceous chondrites, although superficially exciting, could suggest a more haphazard non-biological origin.

It's a clever argument, but the whole pack of cards collapses if it turns out that the PAHs are merely the product of earthly contamination. Unlike the carbonate globules, which show at least some evidence of great antiquity, the dating of the PAHs remains wide open to question. Worse still, similar molecules can be found, at least in very minute quantities, within the Antarctic ice sheets. The trouble with PAHs is that they are easily produced by terrestrial processes, and these have to be sifted out of the analysis. The presence of organic molecules in ALH84001 becomes remarkable only if it can be *proved* that they are very old, and that they originate from Mars.

McKay and his colleagues have an answer to this problem. They suggest that because the PAHs are concentrated deep within the meteorite, and become scarcer near the surface, they could not have crept in from the surrounding Antarctic ice. Had they done so, then the pattern would have been reversed, with the outside of the meteorite heavily contaminated and the inner

THE EXPLORATION OF MARS

areas relatively unaffected. However, this argument cuts little ice with the critics. Derek Sears, editor of the journal *Meteoritics and Planetary Science*, is particularly scathing. 'These arguments are flaky and simplistic. Weathering is a sloppy process. Things leach in, then leach out. They do not do the obvious.'

THE ISOTOPE RATIOS

British scientists investigated the carbon isotope ratios within the PAH molecules. Monica Grady of the Natural History Museum in London said in August 1996, 'We discovered a lot of organic carbon, highly enriched in the [lighter] carbon-12 isotope.'

If the British team's findings do indeed indicate a bias in the PAHs in favour of the lightweight carbon isotope, then this would favour a biological origin. However, Grady adds a note of caution, warning that terrestrial contaminants show a similar bias. 'Even if it can be proved that the PAHs in the meteorite *are* biological in origin, there is no obvious way of telling which planet they were created on until we get more data about the Martian environment.'

There are many complications with the isotope analysis. All we have to go on are isotope samples of the Martian atmosphere dating from the Viking missions of 1976. There is no certainty about the kind of isotopic environment that may have prevailed three or four billion years ago. What's more, there is only the behaviour of terrestrial biochemistry against which to compare the meteorite samples. There is no proof that ancient Martian organisms might have shown the same bias towards one carbon isotope over another as observed in terrestrial biology.

Sophisticated though they may be, the modern isotope analysis techniques are still too far removed from an understanding of the original chemistry which forged all the mineral features within ALH84001. The scientists are working at a great remove of time, space, decay and historical uncertainty. They cannot be absolutely sure of their findings. Even so, none of the isotopic analysis so far performed on samples of the meteorite has actually ruled out the possibility of life chemistry.

Richard Zare of Stanford University is now preparing to make a more detailed examination of the carbon-12 and carbon-

13 ratios within the meteorite's PAHs. Critics of the McKay team's paper have suggested that a detailed isotopic investigation should have been completed thoroughly before any of the team's claims were published. However, Zare suspects that the PAHs are indeed derived from biology. 'In conjunction with all the other data, it seems most likely to me that they all came from the breakdown products of something that was once alive.'

It may be that the isotopic data from ALH84001 will never prove conclusive. However, in November 1996, the British team working on the investigation announced results in connection with another of the SNCs: EETA79001, the same rock which, in 1979, proved that all the SNCs come from Mars.

AN ANALYTICAL ARTEFACT

Colin Pillinger, and his colleagues at the Open University in Milton Keynes, narrowly missed garnering the glory for finding traces of 'life' in a rock from Mars. Shortly before Richard McKay decided to produce a paper for *Science*, he alerted Pillinger. 'There'll be some good news for you.' Pillinger's team would be credited with valuable preparatory work in McKay and Everett's paper.

As long ago as 1989, Pillinger and his colleagues Monica Grady and Ian Wright published a paper in *Nature* magazine, outlining their discovery of organic compounds within EETA79001. As the team was at pains to point out, they weren't saying they'd found life chemistry. They'd merely identified some PAHs. Because these compounds are commonly associated with carbonaceous chondrules, but not *necessarily* with life, the British scientists steered clear of reaching dramatic conclusions. Nevertheless, it was interesting to come across these organic compounds in a Mars rock.

On October 31, 1996, Pillinger, Wright and Grady held a press conference in London to discuss some further work they had conducted that year on EETA79001. In part, they were motivated by a desire to make specific comparisons between that meteorite and a sample of ALH84001. Their one-page press sheet contained a fascinating claim. 'Today, at a meeting to discuss the possibility of life in the solar system and beyond, hosted by the British Minister of Science, Ian Taylor, and held at the Royal Society, [we] offer the strongest support yet for the hypothesis that life once existed on the planet Mars.'

The British work on ALH84001 had produced a new carbon isotope profile within 'a component of' the carbonate material, 'suggesting that it was formed from microbially produced methane'.

The press release also stated that the PAHs within EETA79001 had taken on a new significance in the light of McKay's findings. Since EETA79001 is one of the *younger* SNCs, and since it appears to have been chipped off the surface not long ago, Pillinger and his colleagues suggested that life on Mars may have existed as recently as 600,000 years ago.

One of the methods adopted by Pillinger, Grady and Wright for their investigation is sharply reminiscent of the pyrolysis techniques used aboard Viking, except that compounds are not simply heated to liberate vapours at different temperatures. They are *burned* in oxygen to create and liberate carbon dioxide.

Quite literally, the PAHs inside ALH84001 are burned like fossil fuels ....

The Stepped Combustion procedure takes advantage of the fact that PAHs and carbonates will burn in oxygen at different temperatures. The scientists need not fear confusing the carbon driven off by the mineral compounds with that of the organics. Furthermore, although oxygen is involved in the procedure, any carbon atoms found in the newly-minted carbon dioxide still has to come from the meteorite.

From here on, the isotopic analysis of the carbon dioxide is performed by a mass spectrometer.

The isotope data obtained from this work attracted far less attention in the serious scientific journals than Pillinger's choice to go public with the findings before he presented a paper for scientific peer review. Even McKay and his team had the *Science* paper well in hand before the ALH84001 story leaked out. On November 7, 1996, the journal *Nature* carried a sober report on the London press conference, but their headline for the news item, 'British Scientists Seek Recognition of Role in "Life on Mars" Debate', betrayed the slightest hint of disdain. A subsequent leading article discussed the merits of proper technical peer review, as compared to 'science by press conference'.

Subsequently, Pillinger wrote a letter to *Nature* refuting the criticisms. 'They printed it but left out the crucial paragraph', he says. His letter was outspoken. 'Peer review suppresses debates in print, forces them to take place anonymously in editorial offices, and hinders the publication of non-consensus views.'

After this skirmish, nothing happened for some months. There may have been a certain amount of embarrassment behind the British team's silence. On closer examination of their data, Pillinger and his colleagues found that the isotope ratios were not quite so secure as they had thought. 'I said at the time we shouldn't have gone public with those numbers!', Monica Grady reveals today. 'The initial findings looked very good, but subsequent to October 1996 we did a lot more work. This was such a hot potato, we really felt the need to go over our results very carefully.'

No scientist likes admitting that an experimental result is wrong. In the case of the ALH84001 research, the words 'wrong' and 'right' hardly apply. The data is so delicate, and so difficult to interpret. The laboratory procedures are not easy. Nevertheless, Grady comes close to confessing an error. 'After October 1996, we were happy to conclude that some of our initial isotope fractionations resulted from an analytical artefact.'

YES OR NO? IS IT OR ISN'T IT?

Today, the debate continues. We need more solid data. So far the pro-life explanations to account for the carbonates and PAHs have not yet gained an absolute victory over the non-life possibilities. It remains extremely difficult to prove the antiquity of the PAHs. In addition, any structural analysis of the supposed fossils' interior structures will push against the extreme limits of our technological ability, because the features are so tiny.

These problems do not detract from the achievements of all the scientists who have examined the SNC meteorites in such minute and painstaking detail. If nothing else, they have obliged rival scientists to look more closely at what can happen at microscopic scales, in unusual circumstances, to mineral surfaces.

McKay remains optimistic about the results derived from ALH84001 to date. 'Any one of our findings could be a product of mineral reactions,' he says, 'but the separate indications that we found, all in close proximity, are much harder to account for without a biological explanation. You have to explain all these things together, not just in isolation. In the end, the biological model is just so much more straightforward than all the others.'

The challenge of proof works both ways. Just as the Viking scientist Norman Horowitz was disheartened 20 years ago by his failure to disprove Gilbert Levin's arguments in favour of life, so the scientists who disagree with McKay and Everett have an equally tough time making *their* arguments stick.

LOSING GROUND

Some of the arguments *against* the presence of fossilized life in ALH84001 gained ground during the first months of 1998. Admittedly the 'anti-life' camp got off to an inauspicious start, accusing McKay's team of actually having created the worm-like formations themselves during the atom-by-atom gold plating that precedes electron microscopy. (The scanner's electrons sometimes need a metal surface to bounce off, and in certain cases even the most delicate samples must first be coated.) In *Nature* magazine Everett replied somewhat tersely to these charges that *of course* a wide range of control experiments to test different coating metals had been conducted to eliminate the possibility of an accidental 'analytical artefact'.

Then Adrian J. Brearley at the University of New Mexico asked himself, 'What was the big event that happened to the meteorite in its lifetime? The shock, the shock, the shock.' Brearley believes that the impact of the *other* meteorite which must have knocked ALH84001 off the surface of Mars in the first place would have heated it to more than 500 degrees Celsius, and few people disagree with that estimate. Brearley is prepared to accept that acidic water flowing through the rock during its period on Mars could have created the carbonates, but it was the heat of impact that created the magnetite grains *from* those carbonates. His non-living chemical explanation is realistic, so long as a high temperature (say, 550 degrees Celsius) is included in the theory along with a massive hammer-blow of extra pressure to provide all the heat. Even McKay admits that 'Adrian's idea is quite good.' Scientists are now busy hitting carbonates with high-impact hammers to see what happens.

Remember that Viking took some radioactive carbon-14 to Mars specifically so that it couldn't be confused with any carbon native to the planet when tests were made for signs of organic chemistry? By what seems like the cruellest irony, Timothy Jull at the University of Arizona found significant traces of carbon-14 within ALH84001 and concluded that the meteorite must have been contaminated with earthly products. This remains a moot point, but certainly none of the Viking data supported the existence of carbon-14 on Mars.

The worst news came when Jeffrey L. Bada at the Scripps Institute in California bypassed everyone else's obsession with PAH compounds and went looking for slightly more complex organic chemicals: specifically, amino acids. No one expected aminos from Mars to be associated with *fossilized* microbes. As we noted earlier, the carbonate rosettes and so forth would be far removed from the original life forms, and even the PAH compounds would represent only a decayed remnant of any original biological compounds.

Bada argued that if he could find amino acids in ALH84001, they couldn't be part of any fossils and they would have to be the products of relatively recent terrestrial contamination. Once one allowed for the presence of amino acids from earth in the meteorite, then the PAH data would be suspect also, because, as we have seen, PAH compounds can't easily be dated. Perhaps the PAHs would turn out merely to be *recent* decay products from earthly amino acids?

Having successfully found evidence for amino acids in ALH84001, Bada matched them with similar aminos typically found in the Antarctic ice from where the meteorite was first recovered. ('Pristine' ice is contaminated with environmental fall-out of many kinds.) Bada suggested that the mineral 'fossils' and the organic compounds in the meteorite, while undoubteldy found alongside each other, weren't necessarily related. Reluctantly, McKay said, 'We agree with Bada that there must, after all, have been a fair amount of contamination in this meteorite. It'll make it harder for us to prove that any life signs we find are from Mars.'

Finally, while so many scientists' attentions had been focused upon the isotope ratios for carbon, Harry McSween of Tennessee University examined what might be thought of as the less 'glamorous' chemicals within the suspect carbonate rosettes; in particular, the sulphur. He concluded that the dominant isotope he found, sulphur-34, was too heavy to have been the product of bacteria. Certainly no living thing on earth favours such a weighty form of that chemical. He also found some examples of the magnetite grains growing *out of each other*—a sign that something might be wrong with the purely biological explanation for their presence in the meteorite.

Not that any of this proves anything for good. Bada says, fairly, 'The history of this meteorite is too complex to settle the issue. To answer the question properly, we're going to have to go back to Mars and either analyse the rocks more thoroughly than Viking did, or bring some back.'

As Jull's colleague Dr Warren Beck cautions, 'Even if we do eventually find that the organic material in this meteorite did come from earth, that doesn't rule anything out. A meteorite represents only a tiny fragment of a planet.'

It's not really reasonable to expect more definitive evidence from ALH84001. After all, the meteorite's billions of years old. It's been smashed into by another meteorite. It's drifted through space for millions of years in extreme conditions of vacuum, cold, and exposure to radiation. It slammed into earth's atmosphere at five times the speed of sound; and it's been exposed to 13,000 consecutive winters in one of the coldest and most inhospitable environments in the world: the Antarctic. These circumstances are not suited to the preservation of delicate biological clues.

We have to go to Mars and obtain some fresh samples.

CHAPTER SIX
RETURN TO MARS

WHEN THE VIKING missions came to a close at the end of the 1970s, Carl Sagan suggested that Martian soil clearly demonstrated life-*like* qualities, even if the landers' experimental packages had not proved the existence of life itself. This discovery, he suggested, was almost as good as finding actual organisms.

Twenty years later, Jack Farmer of Nasa's Ames Research Center said much the same thing about ALH84001. 'Even if it turns out that life did not exist in this rock, the data shows that Mars had the basic conditions for life. These are very good reasons for going back to Mars and looking for more evidence.'

The space agency Nasa has determined to do just that.

The loss of Nasa's expensive Mars Observer spacecraft in August 1993 stimulated the development of a series of simple, lightweight and low-cost probe vehicles. At about the same time as McKay's team announced the findings from their research on ALH84001, Nasa administrator Dan Goldin was happy to announce that two Mars spacecraft were undergoing their last few weeks of pre-flight checks prior to launch.

Mars Global Surveyor, MGS, lifted off aboard a Delta II rocket on November 10, 1996, and sped safely on its way to Mars on the correct trajectory. Although a fault developed in the positioning of one of its two solar panels, the majority of its systems seemed to be functioning well.

The MGS carried a number of instruments duplicating those lost aboard the Mars Observer in 1993. In particular, there were cameras, precision radar altimeters and other devices to support an extremely detailed mapping operation, with resolutions of surface features down to a scale of three metres, in essence, the size of a car. (Unfortunately the old Viking landers are too small to spot, although they might betray their presence by the unusual brilliance of the sunlight reflecting from their grey-white surfaces.) MGS's sensitive spectrometers and thermal imaging systems were capable of searching for water just under the planet's surface. In all, the probe should remain operational for five years. Its electronics bay included a subsystem called 'Mars Relay', a communications device intended to support later lander missions down on the surface.

With its total cost of less than $250 million, the entire machine was scarcely more expensive than a Hollywood special effects blockbuster. Cheap, innovative designs allowed Nasa's engineers to make the best use of shrinking budgets. Yet, in some key respects, this reduction of the hardware's traditional sophistication may have added to the complexity of the mission as a whole.

For instance, unlike the old Mariner-Vikings, the MGS could not support large fuel tanks. It carried only half the fuel required to put the ship into the same kind of stable orbit which the Vikings achieved so effortlessly in 1976. When MGS entered the Martian gravity field during September 1997, it fired a small engine to achieve partial braking. This put it into a very elliptical orbit, with a low point of 378km and a giddy high of 56,000km. This was far from the kind of orbit preferred by the Nasa mapmakers. They would have to chart the planet's surface from a haphazard series of scans taken at different altitudes, which was hardly ideal. They wanted MGS to be placed into a circular orbit so that the ship's average height above the ground didn't vary. In order to achieve the best possible visual interpretation of the surface, the mapmakers also decided that no matter what part of the planet MGS happened to be photographing,

RETURN TO MARS

they would like each scan to be initiated two hours after midday, when the sun is high in the Martian sky but just beginning to cast some useful afternoon shadows to reveal surface textures.

The final mapping orbit was supposed to be angled to Mars' equator so that MGS would always fly over new territory while the planet turned beneath. According to the best-case scenario, MGS had to circle Mars once every two hours. Meanwhile, as the mid-afternoon timeline shifted around the Martian globe, the craft would encounter ideal lighting conditions once every orbit and initiate its scans accordingly.

But how to achieve this effect from such an elliptical starting orbit? The mission planners could count on very little help from MGS's feeble rocket motors. They had already calculated that carrying a full load of rocket fuel aboard MGS would have increased its take-off weight by 20 per cent. This in turn would have placed the mission beyond the capabilities of the Delta II expendable launch vehicle allocated for the trip. Nasa's Administrator Dan Goldin established strict budgetary restraints for MGS, and its designers had to work around an inexpensive launch booster of modest power. (The Vikings were carried aloft by heavy Titan rockets whose original use was to orbit two-seater space capsules during the mid-1960s. Modern Deltas are comparatively lightweight machines designed for lifting unmanned earth satellites.)

In the absence of powerful retro rockets, the 'circularization' procedure for MGS was supposed to work like this:

At the lowest point of the probe's egg-shaped orbit, it would drag through the thin outermost layers of the Martian atmosphere, partially slowing down as it skimmed through. This cycle of aerobraking and escape would be repeated continuously over four months, starting at a rate of one 'dip' every 45 hours. Each time the probe encountered the atmosphere it would lose more energy and its orbit would become more rounded.

By December 1997, the vehicle's elliptical path around Mars should have become perfectly circular. But as so often in space exploration, there was a slight problem.

MGS incorporated two flat solar cell arrays, shaped like square paddles. The topmost surfaces were covered with photovoltaic cells, and the opposite sides, facing the direction of travel, were clad with a thin ceramic protective shield. During the 'drag' phase, the two panels were arranged in a 'V' shape, like oars, so that they could act as aerobrakes. As mentioned, one of these panels was deployed a few degrees out of alignment, and when the aerobraking was initiated, the faulty panel began to fold back like a broken wing. Each dip into the atmosphere only seemed to push the panel back further. There was a real risk that it might collapse, unsettling the craft completely.

Eventually it was decided to expend some precious fuel to increase the minimum altitude of the elliptical orbit so that the ship would only skim the very thinnest wisps of atmosphere on each pass. The drawback was that 'circularization' wouldn't be completed for up to a year—if at all.

After three weeks, the aerobraking was temporarily halted with another brief rocket burn. The resulting orbit, while far from circular, was at least less eccentric than before. Making the most of a semi-successful job, MGS was ordered to switch on its cameras and scientific instruments.

Now at last, the benefits of the mission became clear. The camera images were stunning in their clarity, and by April 1998 the scans supported all the current theories that Mars was once a warmer and wetter world. The delicacy of some of the water-cut features, entirely invisible to Viking's cameras in the 1970s, couldn't have been created simply by sudden colossal eruptions of meltwater, because these huge aqueous catastrophes could only have been responsible for equivalently huge and crude features. The fine tracery of channels at smaller scales, and the presence of delicate soil deposits within them, suggested slow, gentle flows sustained over long periods. In May 1998, MGS Project Manager Glenn Cunningham stated with confidence, "You can actually see evidence of water activity in some of these channels that implies recharge by rain from the atmosphere."

Channels cutting across older channels suggested a complex history of water flows rather than a one-off period of flooding. This was very much the kind of news the 'pro-life' lobby wanted to hear. It's not just the presence of water but its availability *over time* that is essential to life.

Meanwhile, the probe's extremely sensitive magnetometers found evidence of anomalies in the surface material of Mars. Countless outcrops of rock seem to have retained some kind of weak magnetic charge from long ago—which is what one might expect to find on a planet that during its early life had a molten core which has since solidified. The magnetic data from MGS doesn't present any glaring contradictions to the negative Mariner and Viking magnetic results. The planet *as a whole* doesn't possess much of a field today but, once again, modern Mars has revealed itself as a deceptively quiet shadow of its ancient self.

Aerobraking for MGS was resumed—cautiously—between November 1997 and March 1998, improving the orbit further. The mapmakers may yet have their wishes granted.

MARS PATHFINDER

After its launch in November 1996 MGS swept across the solar system on a gentle, curving trajectory. The idea was that Mars' gravity field should eventually capture it, but not actually smash it into the surface. That would have been a disaster. A second Nasa probe, the Pathfinder, was counted a tremendous success when it did just that: hit the Martian surface like an artificial meteorite.

Although it was launched a month after MGS, Pathfinder arrived three months earlier. This is because its trajectory was much more direct. When it reached Mars, Pathfinder was never intended to assume orbit, nor was there any time to assess its landing opportunities. The entire vehicle plunged straight into the atmosphere. Touchdown on the surface was scheduled for July 4, 1997, Independence Day. The target zone was the Ares Vallis region, one of those fascinating areas of Mars apparently scarred by ancient water channels.

Pathfinder was clad in an aeroshell rather like a Viking descent capsule, but here the similarity ended. This time, retro rockets played little part in the ship's braking procedures because of the cost and weight implications of carrying sufficient fuel. The final descent was cushioned mainly by parachutes. After the aeroshell fell away, Pathfinder inflated an array of gas balloons, like drivers' air bags, to protect the vehicle against the final shock of touchdown. The underlying philosophy of the ship's design was simplicity, simplicity and more simplicity. Instead of using complicated gears and electric motors to deploy antennas and instruments, Pathfinder fired tiny explosive charges attached to restraining straps. When they snapped apart, the required items of equipment popped out on springs.

On landing, Pathfinder's gas balloons bounced the entire craft at least fifteen times, during which the extraordinary assemblage rolled along the surface for hundreds of metres before finally coming to a halt. Of course there was no one on hand to watch and record all this. Controllers at JPL analysed data from accelerometers inside the craft to derive an approximate version of events.

Pathfinder's first job was to confirm that it had settled more or less on the straight and level. If the craft had come to rest badly tilted to one side or another, or even upside-down, the balloons were supposed to be deflated one at a time in an orchestrated sequence designed to set it upright. All being well, the ship would then deploy three solar panels which would double as physical stabilizers.

The tiny ship was something of an orphan. Unlike Viking, its radio systems had no mothership to act as a relay. To compensate for this shortfall, the transmitters had to locate earth immediately. The aerials first pointed at an obvious target, the sun, and then Pathfinder's computer calculated where the earth was in relation to it. This procedure had to work because there was no other method for making contact.

As it happened, the landing and deployment were flawless, and the ship's radio transmitter quickly reported home. Newspapers around the world celebrated the event as though we'd never reached Mars before. The first colour picture beamed back from the surface made the front cover of a special issue of *Time* magazine for July 4, and all the major papers carried the image on their front pages. NASA's websites logged four million 'hits' a day from all around the world, establishing new Internet records that haven't yet been beaten. But veteran Viking scientist Gerald Soffen couldn't help feeling a sense of

RETURN TO MARS

nostalgia for the old days, even as the world's press heralded the news. He was on hand at JPL to watch the data coming in, and his response to the successful touchdown was tearful as well as joyous: 'I've been waiting twenty years for this moment.'

Too bad that Carl Sagan couldn't have shared in the news. He died in December 1996 at the relatively young age of sixty. Immediately after Pathfinder's arrival, Nasa chief Dan Goldin announced that the landing site would be referred to as the 'Sagan Memorial Station' in his honour.

SOJOURNER

Strapped to the inside of one of Pathfinder's solar panels was a separate vehicle, the Sojourner. It was about the size and shape of a microwave oven, with a roof of solar panels and six little wheels slung underneath its square body. Sojourner possessed a similar bug-like intelligence to the Viking, and could take responsibility for many of its own actions. Its job was to traverse the surface, take photographs and examine rocks.

However there was still a certain amount of 'driving' to do from JPL. Data from the rover's stereo cameras and laser rangefinders was fed into electronic stereo eyepieces back on earth, so that the controllers wearing them could decide where best to send the rover next. A series of instructions was then prepared, tested on a computer simulation of the nearest landscape (based on the stereo camera and laser data), then transmitted to Mars for the rover to obey. Real-time moment-by-moment driving by remote control was, of course, impossible because of the radio time lag; yet it was extraordinary the degree to which humans had achieved a virtual presence on Mars in three-dimensional colour.

Sojourner 'driver' Brian Cooper says, 'The machine was smart and resourceful, and surpassed all our wildest expectations, but it could be petulant at times. Night-time temperatures on Mars were so cold that the rover would fall unconscious. Each morning we had to wake it up with a little heat from its internal batteries, before the sun came up and we could use the solar panels on its roof to generate some energy.' This fragile source of power was far removed from Viking's heavy and

THE EXPLORATION OF MARS

expensive nuclear power packs. At best, JPL expected Pathfinder and Sojourner to last a couple of weeks after touchdown before freezing up.

Sojourner carried an 'Alpha Proton Spectrometer' similar in principle to Viking's X-ray equivalent, only much more compact. (Remember that such instruments can seek only elements, not compounds.) In order for this to function, the little rover had to push itself against a likely-looking rock and then stay in place for several hours while the spectrometer went through its paces. As expected, the dust covering the rocks was found to be rich in iron oxides. It was some of the rocks themselves that turned out to be surprising, with a far higher silicon content than had been expected. In general, they appeared to be far more varied than at the old Viking sites, presumably because ancient flood waters had redeposited them from elsewhere on the planet. (After all, Pathfinder's landing site had been deliberately targeted *because* it was a flood plain.) Many of the larger rocks showed signs of roundedness, as though they had been smoothed and partially eroded by running water, while others

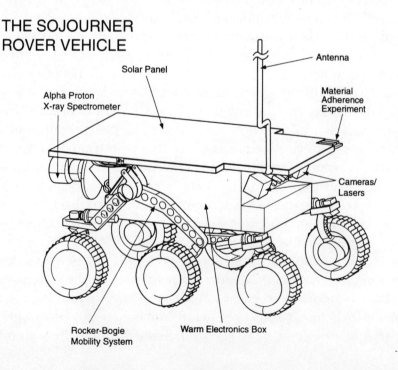

THE SOJOURNER ROVER VEHICLE

seemed to be made up from jumbled aggregates of sand and pebbles—another classic indication of water-driven deposit processes (and precisely the kind of mechanism that can contribute to the production of clays).

Sojourner's proton spectrometer detected silicon at levels higher than expected in several rocks. For a change, iron wasn't necessarily the defining element. The geologists at JPL have speculated that these rocks, at least, were similar in their silicon content to 'andesites', a group of minerals found in the Andes mountains. Since the Andes are a tectonic product, where plates of planetary crust floating on magma push against each other, perhaps Mars had once pushed up mountains or ridges in a similar fashion, albeit less extensively than on earth? The key point is that andesites (and perhaps their equivalents on Mars) cannot be the *first* products of planetary formation, but must result from subsequent geological events.

However, as we have already discussed, few scientists are willing to believe that Mars has been subject to tectonic activity. Apart from the great volcanos, there is a notable absence of high mountain ranges. Mission scientist Harry Golombek explains: 'It's almost like you have two worlds here. One of them shows all this water activity and erosion in the distant past, and maybe tectonic rock products, which we really weren't expecting to find because we thought that Mars' molten interior cooled off a long time ago. So you have a more or less earth-like planet in essence. But then a massive change occurs and you get a much more static environment for the last two, maybe three billion years. Where did all the water go? A good many of us believe it's still there, but under the surface, frozen in various reservoirs. But as for the tectonic evidence—that's a mystery. I don't think there was real tectonism on Mars, even in the distant past.'

So, how can the high silicon content of those rocks be explained? Unfortunately, neither Pathfinder nor Sojourner could make the kind of chemical analysis attempted by Viking, let alone the detailed investigation that many scientists wanted to see conducted in the wake of the ALH84001 affair. Nasa and the associated manufacturing companies regarded both machines as 'technology demonstrators' rather than fully-

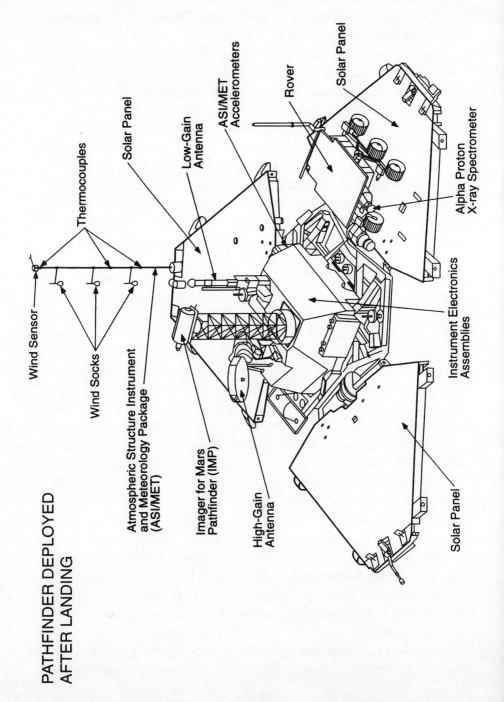

PATHFINDER DEPLOYED
AFTER LANDING

fledged science missions. They were pioneers, nevertheless. Now that that these small, inexpensive devices have proved they can actually work, whole fleets of them can be launched in future years, each carrying more elaborate experiments.

Cheap technology has its advantages. In time, investigating Mars Pathfinder-style may prove so inexpensive that university and college students can participate in an exploration program, sending their own ingenious little machines to the surface. Everything hinged on bringing Pathfinder safely to rest in July 1997, without the usual complex and costly array of retro rockets and landing legs. That success has paved the way for a future series of related vehicles.

Sojourner showed signs of exhaustion after September 7, 1997, and three weeks later Pathfinder also succumbed to the cold; but both vehicles exceeded everyone's expectations.

WHAT NEXT?

A thorough investigation of the Red Planet's environment will require rocks to be broken apart, soil scooped up, and deep excavations made. Much of the chemistry so far encountered on Mars may be largely a superficial result of the ubiquitous dust particles which coat every last scrap of the surface, including individual rocks and pebbles. In this context, some commentators at the American National Research Council have been critical of Nasa's current policy towards Mars. Their assessment is that the Pathfinder and Sojourner possessed 'restricted instruments, a lack of mobility and no equipment to produce images during the landing phase'.

It seems that Nasa cannot win. The space agency tries to keep within the limited budgets allocated to its programs by the politicians, and then it is criticized for building bargain-basement spacecraft that are too cheap. There is a good line of argument to suggest that Nasa chiefs may have brought these problems down upon their own heads with their remorseless enthusiasm for glamorous, vastly expensive but scientifically dubious manned activities in low earth orbit.

Current big spending on astronaut programs has played its part in tightening the purse strings on other, unmanned projects. The International Space Station, due to begin construction in November 1997, has soaked up billions of dollars over its long gestation period, and will cost another $27 billion to complete. Critics point out that a wide variety of sophisticated planetary probes, including proper Viking-style trips to Mars, could have been accomplished with the same money. The space station is a wonderful engineering project, but its true intellectual value remains in doubt. After all, what can be learned from repeating the work already done over the last 36 years—orbiting astronauts in space, but not actually *sending* them anywhere, apart from on a few brief trips to the moon?

Engineers understand very well how a variety of manned missions to Mars could be accomplished with budgets equivalent to those expended on the space station, using components placed into orbit by space shuttles. However, such missions are unlikely to gain political backing unless new data from the upcoming probe missions makes a manned flight seem more compelling. The space station, after all, has been justified largely because of its promised possible benefits to medical research down on earth, with rather less attention focused upon the grander aim of promoting, for its own sake, the goal of human expansion into space. The abstract idealism of American space advocacy groups is not sufficient to justify congressional support for Nasa's budgets.

All of which makes some of the latest conspiracy theories about Mars seem particularly perverse. In 1976 the Viking I orbiter photographed a rock outcrop in the Cydonia region which looked uncannily like a giant human face. Ever since then, a popular myth has done the rounds implicating Nasa in a supposed cover-up. Apparently, the space agency concealed information proving the existence of an advanced civilization on Mars—presumably one capable of producing giant rock sculptures. Against their better judgement, senior MGS managers were obliged to yield to the conspiracy enthusiasts by promising that the probe would return new images of the Cydonia features at the earliest opportunity.

Of course, if the Cydonia 'face' had really turned out to be artificial, then Nasa would have become one of the richest and

most privileged agencies on earth. It could have planned an ambitious manned expedition with little risk of underfunding. The conspiracy theorists must have got their wires crossed.

Sadly, an initial MGS scan of Cydonia, taken in April 1998, leaves little hope for the fantasists.

In the absence of a gigantic expedition to Mars commanded by exultant archaeologists, robot probes will make use of every available launch window from now on, at intervals of 26 months. Nasa would like to attempt a 'sample return' mission in 2004. This flight will call for a complex lander with a rocket-powered ascent stage capable of climbing off the Martian surface and then boosting out of orbit and heading back to earth. Such a mission will send the cost firmly back into the billion-dollar zone which Dan Goldin has been so anxious to avoid.

If the sample return mission of 2004 is approved, the much-maligned space station could be used as a quarantine facility for incoming material, just in case Mars *does* harbour bugs that might be dangerous to humans.

In the shorter term, the Mars Polar Lander (MPL) is scheduled to touch down at a site approximately 1000km from the South Pole on December 3, 1999, after an 11-month flight. This time the probe will be equipped with retro rockets, landing legs, and even a sampling arm rather like Viking, and it will be able to touch down very gently; but as always, its controllers will be hoping that it doesn't land on a hefty rock and topple over. Project engineer John McNamee says, 'Everybody tells me that we won't need a hazard-avoidance system because the polar environment is smooth and benign, with no nasty rocks to bump into. I don't believe a word of it. There are bound to be some gremlins down there.'

High-speed digital cameras on MPL's underside will film its descent. Scientifically this makes sense. For the first time, very precise geographical landmarks can be established so that the controllers will know to the nearest few metres exactly where MPL has landed. However, Nasa's PR departments are also looking forward to some very dramatic footage. Stills images aren't enough any more. Television audiences on earth will want to see some movement and action.

As with Pathfinder, MPL will rely on short-term batteries for power, backed up with solar panels. Environmentalists are vigorously opposed to the rocket launching of radioactive generators such as those used by the Vikings across many months, or even years, of operation. Like Pathfinder, MPL can't expect to stay functional for very long; maybe 90 days at best. In the extreme cold of the southern polar regions, there can be no question of allowing the craft to shut down at night in case it never wakes up. The trick is simply to avoid night altogether.

McNamee explains: 'December 1999 is the late southern spring for Mars, the early part of summer. The sun will be high enough in the sky for 90 days that it won't set during that time, which is similar to what you'll find at our own North Pole at the height of our northern summer. And as long as we have some sunlight for MPL, we can keep it going. Its electrical curent will keep it warm. But after 90 days the sun will go down, the nights will get too long for us, and the power will fade. The lander's going to get really cold, and things will start to fail.'

This time there won't be a rover. As McNamee points out, 'There are two ways to explore the planet. If you're looking at rocks you go horizontally, rock-to-rock. But if you're looking for water, the only direction to go is *down*, and for that you need an arm and a drill.'

As long as the electricity continues to flow, MPL's sample arm will be very powerful (capable of chipping concrete) and will dig trenches up to a metre in depth, looking for polar water and sub-surface soil. This is the first time that we will have taken a look *underneath* the planet's topsoils. Eight miniaturized ovens aboard MPL, in a package called the 'Thermal and Evolved Gas Analyzer', will test the incoming soil and ice samples with a range of instruments: the proton and mass spectrometers and so forth that we have encountered from earlier missions. The panoply of sub-atomic investigation, once so complex for a spacecraft, is now relatively routine and inexpensive.

But there's a much more drastic method for digging even deeper into the ground. Before beginning its atmospheric entry, MPL will let loose a pair of bullet-shaped mini-probes, which will hurtle nose-first towards the surface. On impact they will bury themselves two metres or more deep into the ground, trailing long data cables and aerials behind them. These 'Mars Penetrators' are based on military experience with so-called 'smart' munitions: essentially cannon shells and missiles with simple but extremely rugged computers inside them. Incredibly, these internal electronics can survive the shock of impact at 200 metres per second. Solid-state scientific instruments inside the penetrators will then investigate the water and chemistry of the surrounding soil and ice.

By September 1999 the Mars Climate Observer (MCO) should have gone into orbit, and this craft will act as a communications relay between MPL and earth, backed up by the Mars Relay radio package aboard MGS. The Climate Observer will aerobrake before adopting a circular orbit. With any luck, all the lessons learned from the broken panel on MGS in 1997 will have been well heeded by then.

A RUSSIAN CATASTROPHE

In November 1996, a Russian Proton booster was raised on hydraulic jacks into a vertical position alongside its support tower at the Baikonur launch station in Kazakhstan. The payload consisted of Mars '96, Russia's latest attempt, after so many failures, to make contact with the Red Planet. The mission had been delayed over several years because of budget problems. However, thanks to the new rapport between the West and its old communist rivals, several European and American scientists and research agencies contributed funds and instrumentation to give Mars '96 a fair chance of success.

The benefits were supposed to work both ways. During the 1980s Gilbert Levin argued vigorously that Nasa should fly a new version of his Labelled Release experiment to Mars. He didn't succeed. The American agency wasn't convinced, but Russian mission planners were altogether more accommodating. Their space agency invited Levin to participate on designing experiments for Mars '96.

It wasn't practical to include a robotic arm, pyrolyzing ovens and all the other support mechanisms for an LR test. Levin decided to concentrate instead on the supposed oxidizing characteristics of Martian surface materials. He wanted to investigate whether or not those much-vaunted peroxides actually existed.

Levin's experiment for Mars '96 was elegant in its design. In the Mars Oxidant Experiment (MOX), a probe with a parallel cluster of optical fibres is pressed against a suitable Martian rock or soil surface. Each fibre is coated with a different chemical. Low-powered laser pulses fired along the cables record whether or not the internal reflectivity of the fibre tips alters over time. This yields data about the chemical reactions taking place between the coated fibres and the Martian mineral surfaces.

Levin helped devise this brilliant experiment in conjunction with a dedicated but grossly underfunded Russian team. In fact the Mars '96 vehicle as a whole was packed with innovative experiments from 21 international contributors.

The mothership was designed to send two landers down to different locations. It would also fire a separate pair of dart-like penetrator probes at such steep angles of descent that they would bury themselves in the ground immediately on contact. The penetrators included their own X-ray spectrometers, along with miniature radio transmitters to beam up their findings. The American MGS 'Mars Relay' subsystem was tuned to receive and pass on their signals.

On November 16, 1996 the Proton booster carrying the Mars '96 lifted off. At first, all appeared to go well. Sadly, within an hour, the Russian mission controllers noticed that their radar position signals for the Proton's upper stage were not correct. This unexpended half of the booster was supposed to be in a specific 'parking orbit' around the earth, so that its rocket engines could be relit later, to propel the 6-ton payload away from earth and towards Mars.

The trouble was, the parking orbit was too low. A day after launch, the upper stage, along with Mars '96, fell back to earth, splashing into the South Pacific, just off the coast of Chile, literally sinking all the hopes and plans that went with it.

We can only hope that future probes will have better luck.

CHAPTER SEVEN
LIFE FROM DEEP SPACE

THERE IS YET another class of objects which occasionally slams into our Earth, creating brilliant ionized streaks in the sky. These are the 'meteor showers' which can be seen at regular intervals. But they are not meteorites. They are very fragile, and almost certainly don't consist of stone or iron. They are brittle flecks of dust and ice. They originate not from the asteroid belt, but from comets.

Even if we have not seen any comets in the night sky for some while, we can be sure they have left gossamer-thin ribbons of ice particles in their wake. Each year the earth in its orbit slices through this nebulous plume like an airliner intersecting the old fading exhaust contrails of another airplane which is long gone over the horizon. The little ice fragments of ice and dust burn for a few brilliant seconds as they encounter the upper atmosphere. Then, they vanish. They are as insubstantial as sprites, just fleeting sparks in the sky. Every year astronomers enjoy a particularly well-known display, the 'Perseid Shower', which peaks in intensity around August 12, with faithful regularity. This shower has been traced to the passage of a comet which last swept across the earth's orbit in 1862. Eventually the trail will dissipate, and with it, the beautiful Perseids.

Meteor showers are connected with comets, but we cannot be quite sure where the comets themselves originated. They are among the most peculiar objects in the solar system. It seems likely that they roam a vast spherical zone far beyond the orbits of even the outermost planets, a zone that reaches out a third of the distance to the next nearest stars. This is called the 'Oort Cloud', named after the Dutch astronomer Jan Hendrik Oort, who first predicted its existence in 1950.

Oort based his calculations on the paths of a few hundred known comets with trajectories that bring them close to the sun at intervals ranging from three years to a few centuries. The majority of comets never stray into the central solar system, and do not reach close enough so that the sun can stimulate them into generating their characteristic glowing tails of ion particles and specks of ice and dust. They remain 'dark' and therefore impossible to track. We cannot even be completely sure that the Oort Cloud exists.

The paths taken by observed comets suggest another puzzle. Apart from the tilted orbit of Pluto (a permanently infuriating space oddity, at least until we can despatch a space probe to investigate it), our solar system is concentrated in a fairly flat zone: the very same plane once occupied by the primordial spinning disc. Just as that disc rotated in a certain direction, spinning inwards towards the protostar, so our modern solar system carries much of that same rotational energy to this day, even including the planets' rotations around their own axes.

A number of asteroids have been gravitationally disturbed by close encounters with Mars and Jupiter, grazing very close and whipping around them at speed. These rogue wanderers are exceptions to the rule. The main asteroid belt between Mars and Jupiter still stays broadly on the flat and level.

But not so, the comets. They seem to orbit at all kinds of crazy angles, punching in and out of the 'plane of the ecliptic' like kingfishers darting up and down through the surface of a lake. But why should they have escaped the flat realm occupied by the solar system as a whole? What kind of energy could have disrupted such vast numbers out of the plane? It's possible that comets didn't form out of the primordial disc, but later, when a

THE EXPLORATION OF MARS

young Jupiter-like planet exploded, sending spurts of gaseous material in all directions, which ultimately condensed into icy clumps. The trouble is, nobody can explain why such a planet might have blown apart. Was it through an impact of some kind? If so, it would have to have been gigantic. Far from shattering, a gas world would soak up the shock like a flabby sponge. Mighty Jupiter easily absorbed the punishment of its celebrated run-in with Comet Shoemaker-Levy in July 1994. The planet's pride may have been dented, and the dirty brown stains of the impact remained visible in its atmosphere for many months, but the old gasbag came nowhere near disintegration.

The fact is, we just don't know where comets originated. It has even been speculated that they are not members of the solar system at all, but vastly more ancient entities wandering between the stars. Still, the more favoured theory suggests that they were indeed created out of the same disc which produced the solar system. They may be the frozen remains of countless accretion clumps which never gained enough mass to pull more material into themselves so as to become protoplanets or protostars. Capturing a comet might be like investigating the unfertilized 'egg' of a planet or a solar system that never came into being.

THE 'TUNGUSKA EVENT'

For all their icy delicacy, comets can be forces of destruction. On June 30, 1908, some Tungus farmers in a remote region of Siberia reported a huge fireball which seemed to 'split the sky in two', accompanied by hurricane winds and deafening booms. Around many parts of the world, the night sky glowed unnaturally bright for two days afterwards. Over London, it was bright enough so that you could read a newspaper. However, at the time, few people thought to make any connection between this extraordinary circumstance and the strange stories coming out of Tunguska.

In the pre-revolutionary Tsarist Russian empire, provincial peasants ranked very low on the social scale. The authorities paid little attention to their stories. It was more than two decades before complacent scientists mounted an expedition to investigate the 'Tunguska Event'.

LIFE FROM DEEP SPACE

Much to their astonishment, the investigators found that 2000 square kilometres of forest had been completely flattened. In the central region of the catastrophe, thousands of trees were burned to a crisp. The obvious possibility was that Tunguska had been struck by a huge meteorite, but for all the damage, there wasn't the slightest trace of a crater, let alone any meteorite fragments. However, if an icy *comet* collided with the earth in 1908, then it could have hit the atmosphere at a velocity of maybe 30 kilometres a second. The ice and dust at its core would never have reached the ground. Yet even a modest specimen a few tens of metres in diameter could have produced a huge explosion in the sky, like an airburst nuclear bomb.

Close examination of the impact site in recent years has revealed countless numbers of microscopic diamonds in the topsoil. Nobody is quite sure why, but it may be a phenomenon connected with intense temperatures and pressures. Apart from that, not a trace of Tunguska's destructive visitor has ever been discovered.

LIFE FROM COMETS?

During close approaches to our world, some comets have been observed by space probes. They are irregular lumps of ice and dust ranging in scale from a few metres to tens of kilometres in diameter—all depending on how one defines the 'size' of a comet. The main constituents of the core, the 'nucleus', seem to be water ice, carbon dioxide ice, and silicate dust grains. These loosely compacted snowballs are so delicate that solar radiation shreds their outer layers and creates characteristic glowing tails of ionized particles.

Most intriguing of all, there is evidence of organic molecules, such as ammonia, formaldehyde, methyl cyanide and other hydrocarbons of exactly the kind that may have existed on the primitive earth. These compounds are mainly in the form of ices, and can easily be liberated as volatile vapours should the comet warm up a little when it nears the sun.

The astronomers Fred Hoyle and Chandra Wickramasinghe have developed a theory which states that planets are 'seeded' from space by comets carrying a rich supply of ingredients.

Hoyle says, 'Our hypothesis is that life is a cosmic phenomenon.' Wickramasinghe was among the first astronomers to prove that the interstellar gas clouds pervading the galaxy are rich in organic materials. He believes that comets and dust clouds are intimately related.

The necessary radio and optical observations are not easy to make, because the different types of compounds within dust clouds and cometary tails are greatly intermingled. The effect on telescope observations resembles the way dots on a glossy magazine picture merge into apparently smooth colour surfaces. Yet when the picture is examined under a magnifying glass, separate ink blobs of yellow, magenta, cyan and black can be seen. Dust clouds and cometary tails are so distant that, from our point of view, a similar merging of the many separate chemical 'colours' seems to occur.

The presence of relatively complex organic molecules inside dust clouds and comets is accepted as a fact. However, it's *not* so certain what the exact composition, diversity or maximum complexity of the individual compounds may be.

The relevant observations are made, one at a time, by instruments which pick out different radio or optical frequencies. Gradually an overall impression of the chemistry is pieced together.

There's a widespread belief among scientists that simple organic material within the solar system's primordial cloud may have become incorporated into comets five billion years ago, while the planets were still forming. Eventually some comets collided with the planets, depositing concentrated organic material in the process. This may have 'kick-started' biology, perhaps on several worlds at once.

Scientists agree that comets must have played some role in dumping raw materials onto the earth. The presence of organic chemistry on Mars (if it is ever confirmed) may also be partly or wholly due to cometary strikes. In fact this was one of the arguments raised during the Viking Gas Chromatograph/Mass Spectrometer debates in 1976. Why didn't the machine find *any* organic traces in Martian soil, deposited by comets at least, if not actually generated by native bugs?

Even Norman Horowitz believed that the GC/MS should have found something along these lines.

A better understanding of comets may help clear up a problem with the classic Miller-Urey scenario.

THE METHANE MYSTERY

In a recent general synthesis of their theories, *Our Place in the Cosmos* (1993), Hoyle and Wickramasinghe state that the famous experiments conducted in the 1950s was based on a scientific tautology. The methane pumped into the glass jar had already been derived from a biological source: natural gas. Almost all the methane on earth today is the product of decayed biological material and bacterial metabolism. The only exceptions are the occasional methane derivatives of industrial processes which, in their turn, rely on biologically-derived intermediate catalysts.

Hoyle and Wickramasinghe correctly point out that the automatic presence of methane on the early earth represents a problem in itself. Where did it come from? If it belched out from volcanoes and other geomorphic disturbances, then why can't we find similar non-biological methane sources in the world today?

Methane provided some of the essential materials of life, and then life synthesized methane. It's another instance of *Alice in Wonderland* circular logic, somewhat reminiscent of the protein-DNA conundrum.

Hoyle and Wickramasinghe think that comets must have been the primary source for methane, and a good deal else besides. Among their frozen organic constituents, comets contain ices of methyl cyanide, CH_3CN. Perhaps this material impacted onto the earth at an early stage in its accretion, when the planet was still hot and molten; in that case, it is necessary to understand the mineral and geological processes that could have folded the cometary material into the broiling planetary magma, then split the methyl cyanide, and finally liberated free methane during volcanic outgassing.

We also need to understand the processes which could have synthesized the methyl cyanide and other cometary compounds in the first place.

Hoyle goes much further than this. He says, 'Methane is found in quantity in the atmospheres of Jupiter, Saturn, Uranus and Neptune, where its presence is indicative of bacterial action on a huge scale.' He's right about the methane. There are great quantities associated with the outer planets. As for the presence of bacteria ....

The very fact that Hoyle can make this assertion with a straight face is an indicator of just how problematic the methane question remains. We have to account for its presence *before* life processes came along to synthesize it. The likely explanation is that some kind of ultraviolet reaction in the early solar system dust cloud was responsible, but so far, the exact mechanisms have not been isolated.

Hoyle's assertion that the giant gas planets support vast swarms of bacteria is entertaining, to say the least. In some ways it can only further the general debate by stimulating scientists with opposing viewpoints to try and disprove his thesis. However, Hoyle and Wickramasinghe are on much shakier ground when they suggest that viruses, and even bacteria, might have been deposited on earth, and the planets in general, by cometary impacts. This idea, called 'Panspermia', has attracted wide attention in the wake of the ALH84001 debate. Before looking for some of the major problems with the Panspermia theory, we need to look at it from Hoyle and Wickramasinghe's point of view.

LIFE IN SPACE

Wickramasinghe suggests the following scenario: tiny dust grains of iron, silicates, graphite (a crystallized form of carbon) and other inorganic materials drift through space, occasionally bumping into organic molecules created by ultraviolet interactions. The grains become coated in organics. If they encounter other similarly coated grains, there will be a tendency for them to stick together.

Gradually, clumps will accumulate whose inner recesses are protected against further ultraviolet interaction. Within these protected pockets, the organic molecules have a chance to accrete into more complex forms.

LIFE FROM DEEP SPACE

Writing for the *New Scientist* in April 1977, in support of more formal papers elsewhere, Wickramasinghe asserted that a form of Darwinian evolution takes place within a dust cloud. There is some degree of competition between different varieties of grain clumps, stimulated by their constant creation or destruction in the interstellar environment. 'The simplest self-replicable system involving clumps of inorganic grains glued together with organic polymer coatings becomes most widespread in the galaxy. Finally, we could imagine that the organic polymer films which separate original grains would evolve into biological cell walls.'

Nearly 20 years later, in an interview for the *New Scientist* magazine in August 1996, Wickramasinghe responded to McKay's announcements about ALH84001. 'The reason that primitive life on earth and Mars appear to be similar is that both planets were seeded by similar organisms.'

Hoyle and Wickramasinghe discuss their joint theory in a detailed text, *Our Place in the Cosmos* (1993). They suggest that if any of these grain-organic clusters happen to become part of a planet, they may provide key sources of material for life. Meanwhile, the grains which happen to be accumulated into *comets* will retain any complexity that they might already have achieved, because they will not be subject to the violent temperatures and pressures of planetary formation. 'It would not be unreasonable to suppose that simple life forms developed and flourished within huge aqueous lakes in billions of large comets' *Our Place* concludes.

Hoyle believes that entire viruses, and even bacteria, fall to earth all the time from cometary dust. Recall that our passage around the sun regularly sweeps us through old comet tails. Hoyle establishes some intriguing links between certain outbreaks of disease and the timings of these interceptions.

The medical science of epidemiology involves the statistical study of infections and their spread. Medical practitioners are often puzzled by the strange geographical patterns of disease distribution. One might expect that a virulent infection would ripple outwards from a 'hotspot'. In fact, a great many of the world's major viral and bacterial outbreaks have not conformed to this obvious pattern. A prime example is the notorious influenza pandemic of 1918-19, which killed millions of people worldwide. In May 1976 the *New England Journal of Medicine* examined its lethal geographic spread:

'The disease appeared on the same day in widely separated parts of the world on the one hand, but, on the other, took weeks to spread relatively short distances. It was detected in Boston and Bombay on the same day, but took three weeks to reach New York city, despite the fact that there was considerable [commuter] travel between the two cities.'

Hoyle thinks he knows why.

We don't simply have to take his word as an astronomer that something is amiss with the medical profession's infection statistics. Unusual findings are commonly reported by the doctors themselves. Several curious patterns are detailed in a textbook called *The Common Cold*, written in 1965 by Christopher Andrewes. In one chapter, the book recounts how a Dutch physician, J.J. van Loghem, made a statistical investigation of old medical records, and conducted a postal survey of ordinary citizens, so that he could investigate the epidemiological history of ordinary influenza outbreaks. Van Loghem found that the distribution patterns were complicated, but that certain general isolated trends repeated simultaneously in areas scattered around the country:

'The rise and fall of colds was almost the same in different places. None of this fits in with a person-to-person spread.' Loghem noted that 'such findings are not uncommon; similar things have been reported by workers in the United States.'

Hoyle has his own astronomical explanation. He argues that influenza outbreaks, and a good many other diseases besides, are distributed by cometary material disintegrating in the high atmosphere. By the time the bugs reach the ground, the winds have scattered them across great swathes of territory. Hence the simultaneous appearance of infection across widely separated locations. However, not all of these cometary gifts have been so nasty, Hoyle believes. Fragments of viral material may have added in some way to the development of DNA and evolution on earth. Hoyle suggests that some viral infections may be 'failed' splicings between cometary and terrestrial molecular entities. A few such splicings may have succeeded in the past, contributing to evolutionary processes. The vast majority of the cometary material passes unnoticed into our environment; and only occasionally will a random variety present us with serious health problems.

Now the case *against* Hoyle and Wickramasinghe's scenario.

WHY SPACE? WHY NOT HERE?

The accumulation of organics around dust grains in Wickramasinghe's model sounds rather like the clay-organic mechanism first proposed by Graham Cairns-Smith, but there are a couple of significant differences.

Cairns-Smith's clays derive from a variety of mineralized rocks, eroded by wind and water and then reformed as fairly complex sedimentary material. They have acquired their overall chemical properties partly as a result of the previous terrestrial geothermic processes which helped to create their initial constituents. A similar situation would apply to the montmorillonite clays proposed by Carl Sagan in 1976 as a possible catalysing agent on Mars. Assuming that Sagan was correct, these would have been eroded and deposited in a similar manner to clays on earth. The river-like features observed by the Nasa space probes certainly support the idea of widescale sedimentation.

By comparison, Wickramasinghe's interstellar dust grains would provide only relatively simple catalytic surfaces. His model of grain-organic accretion is not as potent as Cairns-Smith's theory for clays. This is especially true when it comes to stringing aminos into the protein-like assemblies required for life. In addition, clays and liquid water interact to create a rich variety of ionization phenomena on the clays' crystal surfaces.

Water, at least in its liquid state, requires a home world to keep it warm, and a surrounding atmosphere to prevent evaporation. A dust cloud close to absolute zero in temperature and absolute vacuum in pressure will not support flowing water. The presence of liquid water in wet clay greatly increases the opportunities for organic molecules to create the delicate, sprawling three-dimensional arrays required by life.

So far, clay wins hands down over cloud.

A dust cloud is unimaginably nebulous, with individual molecules and flecks of dust and ice separated by a few metres at best, and many kilometres on average. Over immense timescales there would be collisions and accretions. Had this not been so, our solar system would never have been born. Vast gravity fields interact with the clouds so that stars and planets can condense out of them, but these are events which occur on the grand scale of solar systems and star clusters. The chemistry of life favours a more intimate microscopic environment. A few thousand square metres of clay at the edges of a mineral-rich river delta, spiced with a handful of Miller-Urey amino acids, could achieve a greater level of chemistry than a billion billion square kilometres of Wickramasinghe's dust cloud. It could also do it in a fraction of the time.

To be fair, the Panspermia scenario is not impossible. Simplified versions of grain-organic interactions have almost certainly contributed in some way to the raw materials of life. There is a cogent challenge in Hoyle and Wickramasinghe's assertion that scientists must identify the precise mechanisms that delivered methane and other important compounds to the early earth. Even so, the Panspermia theory in its more extreme form doesn't really explain the origins of complex organic molecules here on earth.

A vital ingredient is missing from the theory.

THE COMPETITION PROBLEM

The Darwinian-style competition which is supposed to take place among the grain-organics in space poses a fundamental difficulty. Hoyle and Wickramasinghe have missed a key point about evolution. It's a product of competition, and competition implies a prize—and a risk of losing out.

On earth, evolving entities gain access to a necessarily finite set of resources. If competing entities fall behind in the struggle for materials, they become extinct. That's why Cairns-Smith's clays can no longer 'compete' with DNA-generated enzymes in the accumulation and catalysis of organic molecules. The swift, hungry products of modern DNA biology absorb any spare material well ahead of the clays. Drifting organic scraps would represent a limited resource, keenly fought for by organisms engaged in ruthless competition.

In 1978 the ecologist Paul Colinvaux wrote a beautiful and influential essay on the theme of competition among organisms at all scales, from the microscopic to the gigantic. It was entitled *Why Big Fierce Animals Are Rare*.

To take one example from Colinvaux's thesis: he examined the relationship between lions and zebras, and found that a very large zebra population is needed to support relatively few lions. Why? Because after the lions have dined, there must be plenty of zebras left alive to breed and provide meals in the future. Since the lions eat every couple of days, and tend to focus their hunting upon inexperienced baby zebras whenever they can be found, the supporting breeding population of zebras needs to be exponentially larger than the lion population.

The same subtleties of population interaction and gradation apply right across the plant and animal kingdoms. Resources are finite, and delicate balances between the consumer and the consumed tend to emerge. This biological balance has a very long history.

The question of energy transfer at the molecular scale also plays a major role. To take another amusing example from Colinvaux's essay: by examining the energy yield of organic material at the fundamental molecular scale of sugars, starches and carbohydrates, he showed that the notorious dinosaur *Tyrannosaurus rex* could not have been a fast-running hunter, as so thrillingly depicted in various dinosaur adventure movies. No amount of food could have fuelled such a bulky beast for rapid pursuits. Colinvaux surmised that this monster must have been a lowly scavenger among the dead and dying. It may not have been a particularly noble way of life, but it must at least have been *efficient*.

Another example of energy efficiency in biological systems forced to live in a finite world is provided by very large animals, such as elephants and whales. They must go to considerable lengths to eat vast tonnages of food, but their great size means that they need expend no energy at all in avoiding predators, none of which are large enough to tackle them. The saving in one activity frees up energy for the other. The largest whales are so enormous they have to be careful in case they expend more energy in feeding than they gain from the intake. So they simply swim along with their mouths open, sifting plankton as they go, making as little effort as possible. Plankton provide an easy diet. These microscopic floating life forms don't fight back and waste a predator's energies. However, they *do* exist in vast, rapidly replenishing shoals, thus yielding an excellent food source.

Energy efficiencies apply with equal rigour to tiny biological entities as well as to the behemoths feeding upon them. The plankton specks consist of microscopic plant species, and vast numbers of fish eggs, shrimp larvae and other animal progeny. The 'parent' organisms waste absolutely no energy on protecting individual offspring. Their spare energy capacity is diverted into producing a great *many* offspring. It may seem profligate to sprinkle countless billions of eggs, larvae and plant cells into the sea currents, only to have most of them eaten; but the energy balance remains resolutely efficient. It has to be, otherwise the plankton-producing species would become extinct.

At all levels, evolution fits individual organisms to balance the benefits of energy gained against the costs of energy lost, in an ever-changing but always finite environment. (Think of organisms that live in really small 'niche' environments.) Tiny entities like bacteria, and even viruses, live and evolve within the margins of these constraints. Biology *is* energy management.

Panspermia posits the creation in deep space of organisms at least as complex as viruses and bacteria. For the theory to work, some kind of energy management should apply to the complex entities evolving in an interstellar dust cloud.

However, at the molecular scale of life chemistry, a cloud's resources are effectively limitless. The grain-organic entities have no need to become 'better' at hoovering up material before rival entities beat them to it. There is plenty to go around, and no imperative to compete. The only real threats to the survival of a grain clump are ultraviolet radiation, cold, vacuum and aridity. These environmental factors remain broadly similar across great spans of time and space. Once a haphazard grain-organic has obtained physical protection against the ultraviolet, there is no particular evolutionary pressure on the assembly to become more complicated. Simple molecules tucked safely away behind the grains will survive perfectly well.

This is the key point: if grain-organics in a dust cloud can afford to be so random in their uses of energy, then there's no biology in operation; for the complex grain surviving *no better* than a simple grain is just wasting energy. This is fundamentally at odds with any biological system ever encountered, right down to the level of viruses.

Cairns-Smith's clay theory demonstrates the value of finite resources in the creation of life chemistry. His clay-organics are 'competing' in a world where great environmental changes can occur over small scales. Pockets of constrained environments were available on the primitive earth, with borders defined by oversoaking or aridity, salinity or acidity, wind or calm, hot or cold. These limits would have allowed sophisticated managers of molecular energy to fare better than poor managers.

In fact this is another good reason why life probably didn't start out in the sea, but spread into it later. In certain respects, the early ocean was too much like a cosmic dust cloud. Miller-Urey aminos were under no more pressure to acquire greater sophistication than Panspermian grain-organics. Today the earth's seas are rich and variegated, layered and complex. Oceanic plants and creatures thrive precisely because they now exist within constraints.

There is a four-stage parallel between the lack of constraints in a Panspermian dust cloud and what actually happened on the early earth.

1) *Creative Constraint:* Four billion years ago, suitable catalytic clay surfaces covered only a small fraction of the earth's surface. They provided opportunities *and* constraints for pre-biological and near-biological entities. Primitive life took hold rapidly.

2) *A 'Wet' Panspermia Cloud:* The spread of blue-green algae into the vast oceans introduced them into what was, at that time, an effectively limitless environment. This explains the immense

period of three billion years or more during which evolution proceeded only very slowly.

3) *The Reintroduction of Constraint:* The gradual change in the earth's atmosphere, caused by the blue-green algae themselves, imposed new constraints. Evolution progressed once again.

4) *Massive Constraint:* The introduction of predation, around one billion years ago, added a fast-acting system of constraint. Within 400 million years, evolution was progressing at an exponential rate, with huge increases in the complexity and ability of all classes of organisms.

The recent evolution of our species shows the creative potential of constraint. In regions such as the plains of North America, or the richly-endowed central areas of Africa, prey species were so numerous that the working methods of the human inhabitants barely developed beyond the level of skilled hunting. By contrast, in more challenging regions like Greece or ancient Italy, where life was not nearly so easy, sophisticated technological civilizations evolved to organize the business of survival. In general, the more complex civilizations fared better and expanded more vigorously than the simpler ones.

For competition in a finite environment to make sense at the scales and timeframes characteristic of Panspermia's dust cloud, one must think in terms of entities the size of protostars and protoplanets: entitities that are large enough to make some impact on available resources, and whose intakes and outputs of energy are of genuine significance in a cloud.

Few people would say that stars are alive or evolving in any way (although some cosmologists have suggested that they might be, and we'll return to this idea later). Wickramasinghe's little dust grains are about the size one would normally expect of a primitive biological entity, but they aren't big enough to turn an interstellar cloud into a finite resource.

The Panspermian grains cannot reach towards life.

Wickramasinghe and Hoyle say that 'the simplest self-replicable system involving clumps of inorganic grains glued together with organic polymer coatings becomes most widespread in the galaxy'. They are correct in identifying precise self-replication as the key to life. However, this is where the Panspermia theory really starts to go round in circles.

REPLICATING THE DESIGN DATA

We have seen how the recent Cambrian Explosion of countless new forms of life 600 million years ago was driven by a sudden and fantastically rapid 'arms race' between eater and eaten. The DNA molecule, and the genetic mechanism in general, allowed successful organisms to pass down winning characteristics to their descendants.

The origin of this kind of chemical self-replication, this method of passing design *information* to successive generations, constitutes one of the deepest problems for molecular biology. We have come full-circle to the protein-DNA dilemma. The dust cloud scenario offers few new insights.

Hoyle describes his comet-borne viral visitors as 'simple'. He says, quite properly, that viruses will hijack the DNA in bacterial cells and subvert them, yet he ignores a significant problem. Viral DNA (or the slightly simpler RNA) is scarcely less complex than its counterpart in bacteria and other more advanced organisms. Panspermia cannot account for the catalysis of DNA from simple ingredients, let alone the 'frozen eggs of simple metazoan (multicellular) creatures' inside swarms of meteorites and comets which, Hoyle suggests, helped seed the earth's Cambrian Explosion.

The idea that viruses arrived from space solves nothing. Panspermia is not impossible, but it doesn't address any of the real questions about the beginnings of life.

PARTIAL PANSPERMIA

That said, live or dormant bacteria could probably survive a trip through space in the deeper recesses of meteorites. ALH84001 has opened up an intriguing possibility in this regard. Ancient meteorite strikes may have despatched living organisms from Mars to earth. We discussed that a solar gravity gradient and other factors militate against the passage of rocks to the earth

THE EXPLORATION OF MARS

from Mercury and Venus, but sending fragments from here to Mars is not quite so difficult.

Earth and Mars were both heavily bombarded by meteorites until about four billion years ago, while the solar system was still settling down after its formation. This is one of the reasons why scientists were surprised to find possible isotopic evidence for life 3.8 billion years ago in the Itsaq rocks of Greenland. They believe that the long period of bombardment must have made the earth incapable of supporting life. Itsaq seems to show that life took hold at the very first opportunity after the impact rates eased off.

This swift appearance of life might be explained by some kind of partial Panspermia phenomenon. For all we know, life may have taken hold on Mars first, then arrived here. Or else a doomed planetesimal supported water, warmth and life for a while, before it was shattered into a swarm of asteroids and tell-tale carbonaceous chondrite meteorites. Maybe 3.8 billion years ago those chondrites actually contained something alive?

One thing seems sure, at least in this author's opinion. The first stirrings of life must have taken place on a planetary body of some kind. Hoyle and Wickramasinghe's frozen comets and nebulous dust clouds aren't sufficiently favourable venues.

However, there are other places in the solar system, large bodies *apart* from Mars, where life might have taken hold. We have focused so much upon the main planets, we've hardly mentioned their satellite worlds.

We can ignore the moons of Mercury and Venus. There aren't any .... Next comes our moon, which we know to be lifeless; although, in 1994, a tiny space probe called 'Clementine' detected ice water beneath the topmost layers of dust in the Schrötinger crater, a huge impact depression just alongside the moon's south pole. The walls of the crater plunge for 3km. The sun rarely rises over the rim. Large areas of the crater floor have remained in shadow for millions of years. Clementine's scientists believe that the Schrötinger crater contains frozen water from an ancient cometary impact, sheltered from the sun's heat by the crater walls. It may be that a closer examination of the site will reveal organic molecules bound up among the dust and ice.

LIFE FROM DEEP SPACE

As for the Martian moons, Phobos and Deimos: they are tiny and jagged, probably captured asteroids.

Jupiter has 16 moons, all of which display a wide range of geological features. The surface of Io, the third largest moon, is in permanent volcanic disarray. The tidal forces generated by Jupiter's gravity are so great that Io's crust can never settle. The crusts of Ganymede, the largest moon, and Callisto, the second largest, are less turbulent, but the patterns frozen into place on these worlds are quite extraordinary.

Neither these, nor most of Jupiter's other moons are suited for life—but there is an exception: Europa, the fourth largest moon. In August 1996 the 'Galileo' space probe beamed back detailed images of Europa's surface. The crust is cracked across the entire moon. Geologists believe that the surface melts and reconstitutes itself at intervals.

A close-up view of a Europan crustal fracture is intriguing. There are parallel strips, with brighter strips running alongside darker ones. It seems as though some dark, slushy material is oozing up through the cracks, leaving a stain.

The Nasa scientists believe that the crust is a thick layer of ice. Geologist Ronald Greeley describes the patterns as 'just like ice floes on the earth's polar seas'. He and other scientists believe that a huge ocean of fluid water, 50km deep, exists just beneath the ice layer. Tidal forces from Jupiter prevent the water from freezing. Quite possibly, the water is comparatively warm. When it pushes up between the cracks and freezes into place, it deposits grains of darker material. What that material might be, no one can say.

Initial spectral observations of Europan features, taken from the Galileo in August 1996 and February 1997, were intriguing but inconclusive. The ship's first fly-by brought it to within 155,000km of the surface. However, this wasn't close enough to make a properly detailed examination of the narrow fracture stains. In November 1997, Galileo's looping trajectory through the Saturnian system brought it closer to Europa once again. There were better opportunities then, because the closest approach was less than 600km. Europa may provide a fascinating contradiction: a small world with no atmosphere, five times further away from the sun than the earth, at minus 145 degrees Celsius, colder on its surface than Mars, yet capable, perhaps, of supporting warm fluid water.

Now the moons of Saturn. There are at least 18, and most of them are very small. The exception is Titan, an earth-sized world. It has an atmosphere of nitrogen and methane 50 per cent denser than the earth's. Water ice is plentiful. Titan has all the ingredients for life, save one. Warmth. Taken overall, this is a very cold world. It may have oceans of liquid methane. However, Titan is such a substantial body that it might support geothermal forces. There could be pockets where life can take hold. They may be scattered, but these niche environments would be no *less* likely to support life than the hidden recesses of modern Mars.

Better information should become available within the next few years. The 'Cassini' space probe, one of the last of Nasa's heavy multi-billion-dollar flagship probes, underwent final preparations for launch in October 1997, after a decade of planning and construction. Cassini's primary mission will be to observe Saturn and its moons from space, including Titan.

Cassini's outward journey will take seven years. In that time it will loop once past the earth, twice past Venus, and once around Jupiter, in a complex series of fly-bys designed to boost its velocity and hurl it towards distant Saturn.

Cassini carries a passenger, a European-built capsule called 'Huygens'. On November 27, 2004, this independent vehicle will touch down on the surface of Titan.

If we encounter any biology on Titan, or anywhere else in the solar system, we may find that it shares some molecular characteristics with terrestrial life. We might conclude that earth life and this 'other' life were seeded from a common source. Alternatively, the 'carbon chauvinists' among us might argue that life couldn't have proceeded any other way.

If we find something that is markedly different from known terrestrial examples, then we can be sure that life will develop on different worlds, adopting different strategies.

If we find nothing at all, then we will be disappointed. We *want* to find life in the cosmos.

LIFE FROM DEEP SPACE

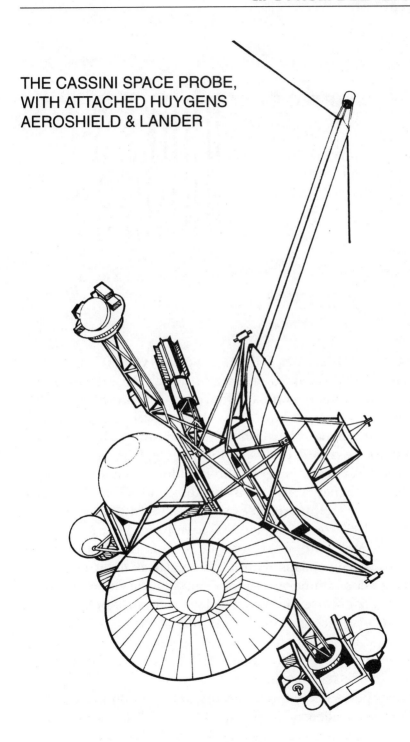

THE CASSINI SPACE PROBE, WITH ATTACHED HUYGENS AEROSHIELD & LANDER

CHAPTER EIGHT
DRAKE'S NUMBERS

IN 1967, JOCELYN BELL, a young Cambridge radio astronomer, detected a regular and fast-cycling signal coming from an extremely compact source in the sky. She and her senior colleague, Anthony Hewish, scribbled a light-hearted note on the paper chart where the repeating blips of signal strength were recorded: 'LGM!' Little Green Men.

Bell never really believed she had discovered signs of alien life, but the signals were unusually rapid, and absolutely regular. Day after day, they pulsed at exact intervals of 1.3373 seconds.

Since 1967, over 300 'Pulsars' have been discovered within the Milky Way, with short pulse rates. Pulsars are stars whose inner cores have collapsed into a dense mass of neutrons, maybe as the result of a supernova explosion. A typical pulsar can be as small as a kilometre in diameter. It spins around its axis in just fractions of a second, sending out intense beams of radiation, like a lighthouse gone mad. The outer layers of a pulsar rotate at a third of the speed of light. An intensely powerful gravity field holds the star together.

Bell's pulsar served as a warning to those with preconceived notions about alien radio signals. Dennis Gilliam, a radio electronics expert at the Lockheed company in California, says, 'As far as artificial pulses are concerned, we can't possibly know what to look for. Take the first time they found a pulsar. Just because it walked like a duck and talked like a duck and looked like a duck, that didn't mean it was a duck!'

We don't even know whether or not aliens would use radio technology. However, it's our only sensible line of enquiry at the moment. So believes an American astronomer, Frank Drake. In 1960, he co-founded a radio astronomy project called SETI, the Search for Extra-Terrestrial Intelligence. Drake's philosophy for managing such a search has applied ever since.

In 1961 Drake decided that SETI's chances for success had to be phrased as a statistical probability. If he could speak to sceptical fellow scientists in a formal mathematical language, instead of on the level of science fiction, then perhaps the SETI concept would be properly considered by congressional funding committees. Drake created a set of numbers, the 'Greenbank Equation', named in honour of the observatory where he was working at that time.

Drake's principle is still used today in discussions about the prospects for finding an alien intelligence. Some people think that Drake's numbers must be a pretty fair reflection of reality. Others think they're hopelessly wrong. Drake himself isn't all that concerned. 'Of *course* we can't know if the numbers are right or wrong. The point is to give everybody a valid framework for discussion.'

THE GREENBANK EQUATION

Drake's formula takes into account the rate of star formation in the galaxy; the numbers of sun-like stars; the likely availability of planets with stable environments; the odds for the emergence of life; the possibilities of intelligence, and the risks that aliens might destroy themselves once they develop nuclear fission, or some other powerful and potentially destructive technology. Along with many other scientists, Drake assumes that technology implies power—and power, a capacity for disaster.

Technological societies are the only ones which are relevant to the Greenbank scenario. An alien race lacking technology would not be capable of sending signals through deep space.

The Greenbank Equation is not an alarmingly complicated piece of mathematics. In its simplest form, it works like this:

Take an italicized letter, f, as an arbitrary symbol, and assign it the meaning: 'the fraction of ...'. Suppose there are a total of ten planets in a solar system, and just one of them supports any intelligent life. Then f (planets with intelligence in that system) would derive a number, 0.1. For convenience, the phrasing in brackets is reduced by assigning some more letter symbols:

N = the total *Number* of stars in the entire galaxy: roughly 100 billion. (N = 100,000,000,000)

P = the total number of stars capable of supporting *Planets* in stable orbits: approximately one in three. (f (P) = 0.33)

E = the total number of planets in a 'P'-type system orbiting within a temperature range suited to the *Existence* of life. With earth and Mars as approximate examples, Drake suggests that two planets might be suited. (f (E) = 2)

L = the total number of 'E'-type planets where *Life* actually arises. Say only one in three. Here, though, the figures become increasingly hypothetical. (f (L) = .33)

I = the total number of 'L'-type planets that have given rise to *Intelligence*. Here the numbers are completely arbitrary, but a figure of one in a hundred seems reasonable. (f (I) = 0.01)

T = the total number of 'I'-type planets which have produced intelligent beings with *Technological* capabilities. Say, one in a hundred. Bear in mind that humans have been 'intelligent' for millions of years, but 'technological' in the astronomic sense for only a few decades. (f (T) = 0.01)

R = the number of 'T'-type civilizations which might achieve *Radio* communication with us, or otherwise signal their presence in the cosmos. At this point, the numbers must be shaved down ruthlessly. Humans have been exploring space for only a few decades, and radio was invented just a century ago. The 'R' number has to take into account the unlikeliness of two separate civilizations, ours and 'theirs', developing technological abilities at *exactly* the same time, and choosing similar means of communication. The 'R' factor is a key point. The galaxy may be teeming with civilizations, and they too may be searching for 'alien' life—us, for instance. Perhaps they have transmitted deliberate signals? If they sent them 200 years ago, or so, then on the cosmic scale of time, their communications technology and ours might be regarded as emerging simultaneously. However, that tiny discrepancy of two centuries could make all the difference between our receiving the signal on earth and missing it. The time taken by radio signals to cross interstellar space adds a complicating factor of anywhere between four years and many thousands, depending on the locations of individual stars within the galaxy.

The 'R' number has to be very small—no more than one in a thousand. ($f(R) = 0.001$)

Finally, the symbol '!' represents the grand prize: the total number of artificial radio targets that we might realistically hope to look for in the sky.

An f fraction written alongside another f fraction means a fraction multiplied by a fraction, and so on.

The Greenbank Equation now looks like this:

$Nf(P)f(E)f(L)f(I)f(T)f(R) = !$

In numerical form, it comes out like this:

100 billion x 0.33 x 2 x 0.33 x 0.01 x 0.01 x 0.001 = 2200

Even though this interpretation of the Greenbank Equation reduces the options to within minute fractions of minute fractions of the total galaxy, it still yields more than 2000 possible targets. The f numbers used here are arbitrary estimates, at variance with the commonly accepted versions of the Greenbank Equation. Opinions differ about the rate at which life may be taking hold in the galaxy at large—if it has done so at all. Conventional interpretations of the Greenbank numbers yield '!' values ranging between 200,000 and 1 million.

Given that there are 100 billion stars in the galaxy, a 200,000 value for '!' might suggest an average distribution of just one civilization for every two million stars. Taking into account the physical shape and scale of the galaxy, and presuming a fairly even spread of civilizations, that would mean the nearest likely target should be roughly 250 light years away from earth. In other words, a radio signal travelling at nearly 300,000km per second, the speed of light, would take 250 years to get from there to here. Immense as it may appear, in the cosmic scheme of things, this is not a very great span of time.

Recent advances in practical astronomical techniques reinforce the value of the Greenbank Equation. It may not be long before some of the f numbers can be filled in on the basis of solid evidence.

OTHER SUNS, OTHER WORLDS

Even the nearest stars are so far away that it's not possible to detect their planets. The general belief that other solar systems must exist is based largely on a framework called the 'Mediocrity Principle'. Despite its rather downbeat name, this is simply a mathematical description of solar system formation which, astronomers believe, should represent a standard model for accretion in dust clouds throughout the galaxy.

The Hubble Space Telescope has observed distant suns in the process of being born. The accretion of material occurs exactly as astronomers have predicted, and the case for the existence of 'alien' planets is hard to refute. The search is on—and over the last two or three years, astronomers have developed a reliable procedure.

A lone star rotates on its axis rather like a planet, but such a star cannot 'orbit' itself in any obvious way. However, numerous stars are members of a 'binary' star system, orbiting around a companion star, often in very brief periods measured in days or hours. If one of these stars is large and bright, and the other is small and weak, astronomers on earth might not be able to see the dimmer star unless they are already primed to look for it.

Even so, they can recognize certain signs. The larger, brighter star will appear to wobble slightly from side to side. Astronomers can infer that its behaviour is being altered by an adjacent mass. The centre of gravity of a binary system as a whole is not in the middle of the big star, but somewhere towards its outer shell. Detailed observations, therefore, will eventually reveal the lesser companion star. However, telescope observations may reveal only one point of light in a suspected binary system, no matter how hard the astronomers search. If the smaller twin isn't a star at all, but a planet, it will be much too dim to make out with a telescope.

Over a dozen candidate stars have been identified in the last two years, with companions that are almost certainly too small and dark to be other stars.

There's one slight *caveat*, however. So far, with one or two notable exceptions, only very large planets with short-period orbits, such as 60 days, have been identified. It takes an object at least the size of Jupiter, which is some 143,000km in diameter, to wobble an adjacent star sufficiently for present instruments to observe it. What's more, the wobble has to be fast if it is to be detected. It simply isn't practical to aim telescopes at individual stars for years at a time.

A short-period orbit is characteristic of a world that is very close to its sun. The normal model for a 'typical' solar system places small, solid worlds in the inner orbits, and massive gas giants further out. The star 51 Pegasi supports a planet larger than Jupiter, but its orbit brings it closer to its star than Venus is to our sun. The object circling 16 Cygni has an eccentric, egg-shaped orbit like that of a careening asteroid, yet its mass is greater than Saturn's. Astronomers believe they may be seeing new phenomena: gas giants spiralling inwards, or colliding; and 'brown dwarfs' that are actually fading stars whose life has been drained away by the gravity fields of larger companions.

There were some encouraging developments during 1996. George Gatewood of the University of Pittsburgh discovered a Jupiter-sized object circling the star Lalande 21185. Its orbital period is approximately 6 earth years, and the gravity equations allow for other smaller planets to be a part of the system as a whole. 'The periodicity and the mass distribution overall looks much more like the sort of thing we'd find in our own back-yard', Gatewood reports.

Lalande 21185 is only six light-years from earth. The Hubble Space Telescope will be pointed in its direction some time in the near future. Within the next decade, techniques will be developed for identifying definite solar systems. Then, astronomers should know where to point their radio telescopes and listen in for signals.

WHERE ARE THEY?

Marvin Minsky is one of the key figures in modern computer research. He's often been called in to advise the SETI astronomers on their data management systems, and on what kind of artificial signals they should look for. Minsky does not require complex logical arguments to refute the Greenbank numbers. His dismissal of the entire concept is legendary in the scientific community:

'If they're out there, why haven't we heard from them?', he says. This barbed comment prompted Frank Drake to reiterate the fact that the Greenbank idea is merely an aid to discussion, not an actual prediction.

The astronomer Frank Tipler suggests that the Greenbank numbers can be applied in a literal sense, as long as only real, not hypothetical numbers are applied. His interpretation plays by Drake's rules in all respects, *except* that the f numbers are not speculative. They are selected on the basis of solid observational data actually achieved by astronomers to date. This version yields an '!' of one: earth.

While Drake and the great majority of scientists in general are convinced that the galaxy must contain at least a few other intelligent species, Tipler is noted for the vigour with which he contests this idea. 'I believe ours is the only intelligence to have arisen in the galaxy.'

Some of the speculations which follow are derived from Tipler's arguments, and from an influential academic paper by Michael Hart, published by the Royal Astronomical Society in 1975, and entitled *An Explanation for the Absence of Extraterrestrials on Earth*.

THE LIKELIHOOD OF SOLITUDE

Conventional Greenbank numbers create their own unusual contradiction. If there really are 200,000 civilizations out there, it doesn't seem too much to hope that out of all those thousands upon thousands of alien cultures, just one of them might have arisen to become a galaxy-spanning civilization. If that were so, astronomers should have detected aliens by now. Alternatively, the aliens should have investigated the solar system.

Our local galaxy, the Milky Way, is approximately 100,000 light-years in diameter. The laws of relativity won't permit a solid entity to travel at the speed of light. However, Nasa's space technologists already understand how an interstellar probe might be accelerated, slowly and gradually over many months, to a velocity of over 15,000km per second, some 5 per cent of light-speed. Given current technology, a space probe would take two million years to reach the far side of the galaxy.

Quite apart from the obvious drawbacks implicit in such a vast span of time, many other problems stand in the way of mounting a *Star Trek*-style trans-galactic expedition. Firstly, the spacecraft would have to be very large, with gigantic fuel tanks, and an expensive nuclear propulsion system. Secondly, humans are not generally capable of planning their activities beyond the average lifespan of an individual—70 years, or so.

Nevertheless, in 1978 the British Interplanetary Society in London published an exact technical specification for 'Daedalus', an interstellar robot probe designed by Alan Bond of the British Aerospace company. Its mission would be to fly to Barnard's Star, the third nearest star to earth, just under six light years away. Since Daedalus would be capable of achieving velocities of 45,000km per second, 15 per cent of light-speed, the journey would take less than 50 years. Daedalus would never return to earth, but it would beam back a mass of radio data immediately upon arrival at the Barnard star system. The data would take six years to reach earth. The project scientists could expect their first results just 56 years after the starship's launch.

In another century or so, it seems certain that humans will be able to design spaceships that could travel at a quarter of light-speed. The closer an object reaches towards the velocity of light, the heavier it becomes. There would be a limit on the maximum speed of a starship.

Given our current scientific assumptions, 25 per cent of light-speed is probably the best that any space explorer could achieve. An advanced manned version of Daedalus could fly to

THE EXPLORATION OF MARS

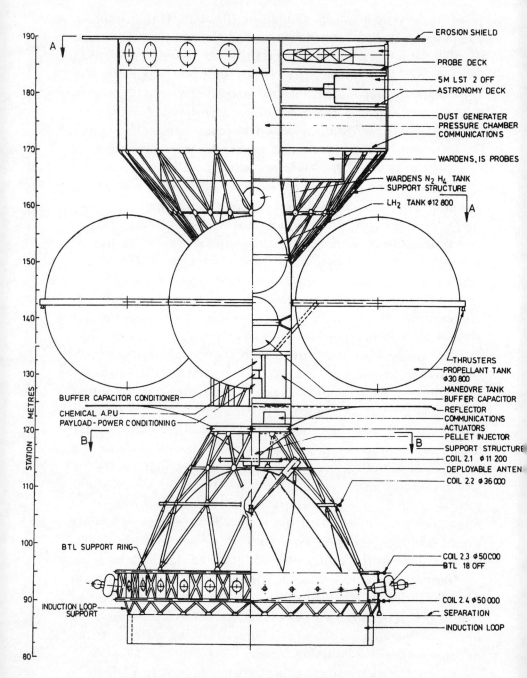

THE DAEDALUS STARSHIP DESIGN

the very nearest stars in only a few decades. Routine traffic among close-knit star systems could become common. However, for human astronauts with lifespans of 70 years or so, several decades might seem too long a time to spend shut away inside a spaceship. Individuals would have to commit their working lives to a mission. This sacrifice might be off-set by building very large, luxurious starships, capable of accommodating substantial communities of people: lovers, families, parents and children.

Another possibility is that astronauts could be placed into 'suspended animation', a kind of cold storage during which their breathing, heartbeats and general metabolisms are reduced to the lowest possible limits consistent with the preservation of life. From the crew's point of view, a 50-year journey might seem to pass in a moment of deep sleep.

Unfortunately, the earth would not be held in cold storage alongside the crew. After a 100-year round-trip, the crew would find that all their relatives back home had aged and died at the normal rate. The starship's outward journey might represent an acceptable emotional challenge to hardened space professionals. The homecoming would be much more difficult.

However, the problem becomes more subtle if the length of the journey is considered as a *proportion* of a given life span, rather than as a specific number of years. Consider how the mariners of previous generations were willing to tolerate sea voyages lasting many months, and sometimes several years. In the late 18th century, Captain James Cook and his crew made a series of exploratory voyages across the Pacific Ocean, with round-trips typically lasting three years. At that time, in the 1760s and 1770s, the average lifespan for an ordinary sailor was no more than 45 years. At least half of that life would have been spent at sea.

Now consider the possibilities of some future century. Medical technology will almost certainly produce great increases in average human lifespans, at least for a privileged minority. If humans can stay alive and healthy for 200 years—barely tripling our current 20th-century maximum—then, for a motivated space explorer, a long starship voyage might seem acceptable, particularly if the tedium and isolation could be eased through companionship or suspended animation. On coming home after a century or more in space, the explorers would have a good chance of seeing their relatives again.

That's just the *human* perspective. It is not unreasonable to suppose that a highly advanced alien culture might have achieved lifespans of many centuries.

GALACTIC CENTRAL

The galaxy's overall distribution of stars becomes denser in its inner regions. Averaging out our Greenbank distribution of 200,000 civilizations, it seems sensible to assume that many of them would be concentrated near the centre by sheer virtue of statistical bias. Perhaps this is where one might expect to find a network of stars linked by the trade routes of alien spacefarers.

Suppose that a proportion of alien civilizations are much more advanced than ours. They should be capable of inventing technologies that make the BIS Daedalus concept look like a school science project. The fact that nobody has visited earth in such machines suggests that there are no space travelling aliens abroad in the galaxy. 'The solar system is an obvious target for investigation, but nobody has come,' Frank Tipler maintains.

It might be that alien cultures simply aren't as interested as we are in space exploration. In an article for *New Scientist* for November 1996, Edward Harrison at the University of Massachusetts advised, 'We should not assume that aliens must share the motivations of humans. Life may evolve to a level beyond imagination, and its motivations may be utterly incomprehensible to us.'

Tipler's counter-argument is persuasive. 'Other civilizations are likely to be diverse, just as ways of life and cultures on earth are diverse. Even if the stay-at-home mentality prevails throughout most of the galaxy, there should be exceptions.' Tipler says that if the Greenbank numbers are even remotely accurate, then at least one of the supposed 200,000 technological civilizations should have opted for a galaxy-spanning way of life. 'It hasn't happened, or we'd know about it by now.' Tipler's final line of argument, developed during 1996, assumes that the Martian meteorite ALH84001 *does* indicate biology on the Red Planet.

This would support the traditional scientific assertion that life gets going fairly quickly and easily wherever a suitable planet exists. If this is so, then the Greenbank numbers should double, or even quadruple. The glaring absence of a galaxy-spanning civilization *anywhere* among a possible million inhabited worlds says that something's amiss with the Greenbank idea as a whole.

Tipler says that everything makes perfect sense if it's assumed that nobody's out there. 'We are totally alone. That makes us special. If we extinguish ourselves, the universe will remain lifeless. All being well, our descendants will spread and colonize the entire universe.'

Harrison says that Tipler's viewpoint is 'very primitive, ungenerous and bereft of imagination'. All the same, Tipler is resolute in his belief that 'SETI is a complete waste of time.' Governmental funding agencies seem to agree with him. The various SETI programs around the world remain severely underfunded.

We listen, as best we can, to the skies. We don't *speak*. An enormous amount of power is required to send a deliberate radio signal to distant stars. The whole exercise is pointless if we don't know where to aim the signal.

One of the intriguing aspects of the pulsar discovered by Jocelyn Bell in 1967 was its omni-directional nature. It was very tempting to imagine that it was a giant broadcast beacon or a navigation aid of some kind; or perhaps an alien greeting which took into account the difficulty of aiming messages at specific targets in the vastness of the cosmos. It's as if the pulsar were behaving like an illuminated signpost in the night, careless of the identities of those who might read it, but wishing to attract attention just the same.

THE DRAWBACKS OF CONTACT

We can't begin to know whether any of the factors influencing the behaviour of our species might apply to alien beings. Just prior to the release of the film *2001: A Space Odyssey*, in April 1968, the film-maker Stanley Kubrick observed, 'Aliens are unknowable to us. Otherwise they would not be alien.' Pointedly, Kubrick's film, co-written by Arthur C. Clarke, did not explain the meaning or behaviour of its central device, an extraterrestrial entity in the shape of a smooth, dark slab.

We cannot guess whether or not an alien would show any political or moral characteristics that we might recognize. We cannot assign wisdom and compassion to aliens simply on the basis of some vague notion of an 'advanced' civilization.

We cannot assume that they are nice people to meet.

The debate on this issue centres on the question of culture shock. Could humanity survive the psychological and social impact of an encounter with a superior species—or, indeed, *any* alien creature? There's another, darker side to this line of enquiry. Would an advanced civilization from another world regard us as worthy galactic companions, or as cattle suitable for exploitation?

Take an unfortunate earthly example: we believe dolphins to be almost as intelligent as humans, yet, in some parts of the world, fishermen will snare dolphins in their nets while pulling in shoals of tuna. Capturing a profitable tonnage of tuna fish is considered more important than the preservation of individual dolphins. That's an example of what one might call 'careless' intervention between different intelligences.

A variation on this theme applies to primate medical research. We acknowledge our kinship with chimpanzees, and regard them as sentient beings worthy of protection and affection; at the same time, we subject them to painful surgical and psychological interventions precisely *because* of their biological and psychological similarity to us. Might alien visitors display similarly contradictory attitudes towards humans?

Another class of intervention might be called 'inadvertent'. When white settlers first encountered the native peoples of Africa, Australia and America, exchanges between the cultures were not necessarily hostile in intent; at least, not to begin with. Yet, great destructions still occurred. After Columbus' discovery of America in 1492, Spaniards flocked across the Atlantic, eager to explore and exploit the 'New World'. They brought with them European strains of bacterial and viral infection, such as smallpox and venereal disease, against which the native Americans had no immunity. The death rates among indigenous peoples were staggering. The westerners were also attacked by unfamiliar diseases, particularly mosquito-transferred malaria.

The 'exploitation' scenario is even more sinister.

Just how civilized must a culture be, before it relinquishes the desire to dominate less developed cultures? Citizens of Europe and America no doubt regard themselves as belonging to the most liberal and advanced nations on earth; yet the Third World nations appear to be suffering the consequences of their exploitation. The relationship between a poor nation and a rich nation is a complex and dynamic brew of 'careless' and 'hostile' interactions.

There seem to be two mild mitigating factors. The current planetary perspective of many diplomats and politicians enables the exploiters to recognize the value of helping, rather than hindering, the exploited. Environmentalists, for instance, recognize that if the rain forests of South America are to be protected for the benefit of the planet as a whole, then regional economic problems have to be addressed. 'Careless' and 'hostile' patterns of intervention are becoming less welcome among sections of the societies principally responsible for generating them.

This trend has a bearing on alien contact. It's possible that very advanced extraterrestrial visitors might be so sophisticated in their moral sensibilities that they'd leave us alone, for fear of wrecking our culture. The *Star Trek* television series supposes that future generations of human space explorers obey a 'Prime Directive' which states that they are not supposed to interfere with the progress of alien civilizations. Interference is posited as a moral affront. Of course, the weekly television show's dramatic narrative requires some degree of interference, but the concepts behind the Prime Directive are quite mature.

If television scriptwriters can dream up such an idea, maybe aliens can too. That might explain the *apparent* absence of their ships in our skies.

As for a 'hostile' intervention: history might suggest that modern liberal democratic nations are less likely to wage war upon other liberal democracies. The world is showing an increasing pattern of democratization on all continents, even if the process is slow. Perhaps this is to the good. Nevertheless, all human cultures co-exist in a constrained, finite environment, no matter how noble their intentions may be. Cultures evolve and compete. That's why they exploit each other. The constraints on the cultural environment are defined just as they are for biological organisms, in terms of territory and energy. Economics and warfare are based entirely around the desire to acquire energy and space at the expense of competitors. No battle in the history of mankind has been fought for any other reasons. No economy is run on any other basis.

What if we were able to get out into an economic equivalent of the Panspermian cosmic dust cloud, so that our resources became effectively infinite? Would we still need to compete?

This real possibility emerges for a space-faring culture. For all practical purposes, the resources of the cosmos are indeed infinite. A very advanced alien culture may be able to extract all its energy from suns, adapt lifeless moons for habitation, spread its peoples to new solar systems, and so dispense with close-fought cultural competition altogether.

If such creatures ever visited us, they would have nothing to gain from exploiting us. No amount of pillage and destruction of this one little planet could match the limitless energies they could derive from the broader cosmos.

Arthur C. Clarke has suggested that any civilizations capable of developing the technology for star travel must, by definition, have solved all their 'energy' problems. Clarke also presumes that social difficulties, too, will have been resolved. The American Apollo missions to the moon clearly demonstrated the need for widescale collaboration on a task so complex as leaving the earth. Interstellar travel must presumably require a proportionally greater degree of cooperation, and that might indicate a civilization at peace with itself.

If aliens visit us, we will find them profoundly cooperative creatures, benevolent and socially at ease. That is, if we actually get to meet them. Clarke's acclaimed novel *Rendezvous with Rama* (1973) posits an encounter with an unoccupied alien craft.

'Rama', a colossal alien cylinder, sweeps through the solar system. A small ship is despatched from earth to investigate. The survey team gains access to the cylinder's interior. Clearly Rama has the capacity to transport thousands of alien creatures, but it appears to be completely abandoned. Rama bypasses earth and heads straight for the sun. The human observers assume, sadly, that it must be off-course, and they try to divert it by attaching rocket motors. Rama relocates its original path, smashes into the sun, and passes out the other side, gaining velocity. Its passage through the solar system was a fuel stop.

The cylinder displays not one shred of interest in humanity. It does not scan our planet. It neither hinders nor encourages the visiting survey team. It doesn't register our presence in any way. Its purpose, utterly mysterious to us, lies elsewhere.

This clever and melancholy story describes a sort of gentle bruising between cultures scraping past each other in the night. It highlights the immensity of the cosmos, and the insignificance of our place within it.

Perhaps we should not rely so heavily on the skies for our inspiration. To quote Stanley Kubrick once more: 'The most terrifying fact about the universe is not that it is hostile, but that it is indifferent; but if we can come to terms with this, and accept the challenges of life within the boundaries of death, then our existence can have genuine meaning. However vast the darkness, we must supply our own light.'

THE LIMITS OF INTELLIGENCE

Alien cultures freed from energy shortages by technology could become stagnant. They might fall into decadence and chaos.

The precedents for stagnation can be observed in historical records. The fall of the Roman Empire is widely regarded as an exemplar. The economic and military power of the Roman state at its height was so great that there were relatively few incentives to strengthen the position further. However, by the 2nd century AD, there were signs of decay. The military support structures had weakened; the economy was corrupted, and the empire fell to smaller armies of conquest from supposedly inferior Hun and Ostrogoth tribes. By 476 AD, the Roman Empire was at an end.

Such a pattern has been typical throughout history: mighty empires became too diffuse, too complacent and unconstrained. Similarly, an alien culture freed from hunger or other driving forces by its own powerful technologies might also collapse in on itself, but without the need for invasion by deliberately hostile forces. Creatures clever enough to design giant starships might become too lazy or too atrophied to build and fly them.

Without constraints, an evolving system cannot progress. Civilizations acquire complexity by adapting to restricted environmental circumstances, just like biological organisms. Where hunting cannot easily provide food, farming civilizations will develop ploughs and other technology to exploit fertile land. Sea-going nations arise to counter the drawbacks of restricted space at home. They focus their efforts upon building ships. Warrior nations prey on weaker nations, settling their lands, or else following a highly flexible, nomadic way of life.

Increasingly more advanced civilizations arise by means of an evolutionary ratchet mechanism. The biologist Richard Dawkins has shown that 'cultural transmission [of advantageous characteristics] is analagous to genetic transmission. Although basically conservative, it can give rise to a form of evolution.' Dawkins has conceived a cultural analogue for DNA. He calls it a 'meme'. Basically, it's a human belief—any important idea passed down the generations through purposely preserved social heritage: patriotism, religious faith, a code of honour, or a technological skill. The memes intermingle when cultures meet. They affect the general behaviour of an emerging society, for better or for worse.

If a civilization is evolving, it must be subject to limitations. However, these limits might disappear if an alien civilization achieves the level of total, god-like power so often imagined in science fiction and space movies. All the arguments that work against Panspermia within a cosmic dust cloud could apply to a civilization that finds a technological way of accessing infinite cosmic resources and frees itself from constraint.

There's a theory suggesting that *no* civilization in the galaxy at large could progress beyond a certain point. Intelligence and technology will eventually push a species towards atrophy. In 1968 the renowned physicist Freeman Dyson warned that it was unwise to confuse technological advance with genuine progress:

'Intelligence may indeed be some sort of benign influence

creating isolated groups of philosopher-kings far apart in the heavens. Alternatively, intelligence may be nothing more than a cancer of purposeless technological exploitation, sweeping across the galaxy as irresistibly and pointlessly as it has swept across our own planet.'

This is a very sombre idea. However, it remains true that our hopes for a friendly, productive, uplifting encounter with a species from some other part of the cosmos hinge on faith rather than scientific evidence. Intelligence is simply not enough. Power is also insufficient. Unconstrained beings, freed from evolutionary pressures, and with all of the cosmos at their disposal, will need *wisdom* to guard against the very freedom that might destroy them.

Even today, some prescient commentators find our highly technological Anglo-American society, overstimulated by trivia and understimulated by genuine need, is showing signs of decay. In a renowned appraisal of western civilization, *The Ascent of Man* (1973), Professor Jacob Bronowski observed, 'I am infinitely saddened to find myself suddenly surrounded in the West by a sense of terrible loss of nerve, a retreat from knowledge .... If we do not take the next step in the ascent of man, it will be taken by people elsewhere, in Africa, in China .... Should I feel sad? No, humanity has a right to change its colour.' Yet Bronowski could not help but mourn the passing of 'the particular civilization that nurtured me'.

What might happen to a highly developed alien species with no 'people elsewhere' to take 'the next step in the ascent' if theirs should falter?

THE END OF EVOLUTION

Suppose some advanced alien beings are wise enough to avert decay, and as a result, they prosper and develop indefinitely. Arthur C. Clarke suggests that a successful strain of cultural and technological evolution, if such a thing exists, could have only one outcome—and very few species would attain it. 'A genuinely successful race will be godlike. There's nowhere else for them to go.' Clarke suggests they might turn 'in their infinite loneliness' to little creatures like us, for nothing more than companionship.

They would seek the kinship of consciousness for consciousness, the same kinship that makes us search the heavens in the hope of finding them.

In 1968, during a rare interview, the film maker Stanley Kubrick suggested that 'aliens might seem like gods to millions of less advanced beings in the universe, just as we would appear like gods to an ant that somehow comprehended Man's existence .... If the tendrils of a greatly developed alien consciousness ever brushed men's minds, it is only the hand of God we could grasp as an explanation.'

CHAPTER NINE
AN ORDERED UNIVERSE?

IN 1929, THE ASTRONOMER Edwin Hubble demonstrated that all the galaxies in the observable universe are flying apart from each other. The further away a galaxy is, the faster it recedes. The light from distant galaxies stretches into longer wavelengths at the red end of the spectrum, just as sound waves from a speeding train lengthen and lower their pitch when the train hurtles past. Astronomers have tried to pin down the 'Hubble Constant' of expansion by calculating the 'redshift' of light from distant galaxies.

According to the accepted scientific theory, the cosmos is expanding because it still carries the momentum of a tremendous explosive event called 'Big Bang', which happened in a single instant somewhere between 12 and 16 billion years ago. At that instant, everything in the universe blew outwards from a single dimensionless point. Elaborate mathematical descriptions of energy, gravity and matter interactions have been invented to calculate the sequence of events backwards in time to within fractions of a second of Big Bang itself. Physicists display great confidence in their models.

Recent discoveries by space probes seem to support Big Bang. All regions of the sky emit a faint microwave radiation. It comes from everywhere, yet from nowhere in particular. It's the warm afterglow of the initial explosion. Over a period of four years, COBE, the Cosmic Background Explorer satellite, found uneven patches in the radiation, where tiny fluctuations in the initial fireball expanded to produce distortions in the fabric of space. Such distortions were crucial. Had the Big Bang been perfectly uniform, all the matter in the cosmos would have flown outwards without condensing into galaxies, stars and planets.

Cosmologists have long believed that subatomic disturbances must have interfered with the too-tidy symmetry of Big Bang within the first fractions of a second. Now COBE has found the microwave echo of those vital imperfections in creation. The proponents of Big Bang are delighted.

The further away (the more redshifted) an object is, the older it must be, because it has travelled so much further outwards from the initial point of creation. That mythical point is just a clumsy convenience of human language. Space and time have also expanded as an integral part of the universe, and since all points of the universe are expanding away from each other at equal velocities, there is no 'centre' of the cosmos for us to discover.

Mathematicians are happy to play around with these ideas, folding space into multiple dimensions and arguing that the shortest distance between two places in the cosmos isn't a straight line after all. They aren't so happy, though, with snags in recent telescope data.

Detailed measurements of countless stars show that their light output changes as they age. By these rules, certain clusters of stars within the heart of our own galaxy, the Milky Way, appear to be incredibly ancient at around 16 billion years old. This is odd, because the galaxy is supposed to be only 10 billion years old.

This problem has vexed astronomers for some years. In 1994 the puzzle became even more acute. Wendy Freedman at the Carnegie Observatory in Pasadena used Nasa's orbiting space telescope in an attempt to measure the Hubble Constant more accurately. According to Freedman, redshifts recorded from a wide variety of distant galaxies suggest that Big Bang

AN ORDERED UNIVERSE?

happened somewhere around 12 billion years ago, the most recent date still consistent with the main cosmological theories.

So the problematic stars within our own galaxy burst into life not just before the Milky Way, but four billion years *before* the first instant of Big Bang. Something's wrong somewhere.

There are other problems with the Big Bang scenario. The gravitational attraction of stars and galaxies prevents the cosmos from breaking apart. However, gravity is connected with mass. The universe isn't massive enough to hold itself together. This is not just a minor shortfall in the numbers. As much as 90 per cent of the necessary mass has yet to be accounted for. There may be colossal amounts of material in space which our instruments cannot detect, because it doesn't emit or reflect light or radio waves. This is the so-called 'dark matter' hypothesis. There are gaping holes in the Big Bang equations where the mass values from all this invisible material are supposed to fit.

Astronomers attempt to resolve these puzzles by calculating forwards in time as well as backwards. They are keen to work out what will happen to the universe in the distant future. Will it carry on expanding and fizzle out like the biggest firework of all time? Or will it gradually slow down, halt for a few billion years, and then collapse inwards, creating a Big Crunch? This question disappears if the Big Bang concept is discounted altogether.

Fred Hoyle is a famous champion of the rival 'steady state' theory. This suggests that the expanding universe is constantly replenished from within by new material. There was no beginning, there will be no end. There's no Crunch to come, and there never was a Big Bang.

Physicists such as Stephen Hawking also think the Big Bang model is flawed. Hawking has described a cosmos created from 'superstrings', quantum distortions in the fabric of the vacuum itself, which can spontaneously give rise to matter. Superstrings are thinner than the width of a single atom, but can give rise to immense energies. The universe they create does not start in a pinpoint fireball. It emerges out of what might usually be described as 'nothing'.

Superstrings are clever mathematical inventions, designed to wish away the gravitational and redshift problems of Big Bang.

They are probably the most advanced ideas the human mind has ever conceived. Yet the strings just create an even bigger question. They work only if the laws of quantum physics are somehow embedded in the vacuum. Those laws are very complicated. Certain versions of string theory require space-time to be entangled among 26 dimensions.

The point is this: it doesn't *matter* that some of the theories disagree with each other, or that our knowledge is imperfect. That's only to be expected. All scientists agree that they are observing a universe which contains precise consistencies at every scale, from atoms to galaxies. The universe has order. It has laws. Human scientific theories are reflections of certain aspects of those laws. They are not a complete phrasing.

So, where do those laws come from? For all their failings, mathematical models of the cosmos, from stars and galaxies down to the orbits of simple asteroids, have proved incredibly effective. But *why* should numbers and equations be so useful?

Stephen Hawking has asked, 'What is it that breathes fire into the equations and makes a universe for them to describe?'

Hawking's work is very complex and almost entirely technical. He seems, on the face of it, to be the ultimate scientific reductionist. Yet he has asked, 'Was it all just a lucky chance? That would seem a counsel of despair, a negation of all our hopes of understanding the underlying order of the universe.'

Albert Einstein pinned down the ultimate mystery when he said, 'It is not the laws of the universe that are remarkable, but the fact that there are any laws at all.'

This is where the 'Anthropic Principle' kicks in. In its 'weak' form, it is simply a philosophical query. In its 'strong' form it allows us to investigate the possibility that the cosmos is, as Fred Hoyle calls it, 'a put-up job'.

The creation of a God intent on bringing life into being.

THE ANTHROPIC PRINCIPLE
The Weak Anthropic Principle works like this:

The universe seems suspiciously well-suited for life. We think like this because we happen to exist within that universe. Had the universe turned out to be unsuitable, we would not be

AN ORDERED UNIVERSE?

here now to discuss the problem. The 'Weak' version doesn't really get us very far. The 'Strong' Anthropic Principle is much more interesting. It examines four fundamental forces of nature, and considers what might have happened if they had turned out differently. Everything described in this book can be ascribed to this quartet of interactions:

1) *Gravity*, which pulls matter towards matter.

2) *Electromagnetism*, which provides atoms with their chemical characteristics at close ranges, especially their positively charged protons and negatively-charged electrons.

3) *The Strong Interaction*, which binds the nucleus of an atom.

4) *The Weak Interaction*, by which atoms decay radioactively under certain circumstances.

These four forces differ in strength, but the differences relate in the same way to all atoms in the cosmos. What would happen if any of these forces were to change, even slightly?

Gravity is such a weak force at small scales that it scarcely counts. However, if the exact strength of gravity weakened by the smallest possible degree, the large-scale implications would be dramatic. The galaxies would disintegrate. Suns would die prematurely. There would be no broad galactic gravitational field to pull the resulting debris into new dust clouds. The formation of new suns would be impossible. We wouldn't be here, because the cosmos couldn't have produced more than a single generation of stars.

On the other hand, if gravity were a little *stronger*, then the rate of collisions between stars would be so great that a typical solar system like ours would not survive long enough to produce stable planets and life.

What about electromagnetism? If the balance of the electromagnetic force altered in any way, all known chemistry would disintegrate. The same applies to the Strong Interaction.

As for the Weak Interaction, it's tempting to think that we could do without radiation. However, if the Weak Interaction became slightly less weak, then the stars could not burn and the elements of life could not be manufactured.

These examples are gross simplifications. Any tiny change in the relationship between the four forces would produce not just the minor inconveniences described above, but the dissolution of the cosmos as we know it.

The four forces which make the universe possible can be found within a single atom. Describe the behaviour of an atom in terms of those forces, and the characteristics of stars, planets, and galaxies follows with absolute inevitability. The same applies to biochemistry. Hawking's biographer Kitty Ferguson has observed, 'If something in the cosmos isn't a product of the four forces, then it hasn't happened.'

The Strong Anthropic Principle takes note of the following circumstance: at the same instant our universe was created from a singularity, the laws of physics must also have been created, because the laws (the consistencies we observe) are an emergent property of the universe itself. There is no law-about-laws which dictates that the four fundamental forces had to come out the way they did. Physicists agree that a Big Bang could produce forces slightly different from the ones we actually find in the world. The varieties of poor-quality balances between the forces are billion-fold. Any of these bad balances would have produced a universe incapable of maintaining order and creating life.

How did Big Bang generate a good-quality universe first time round, against such colossal odds?

The Strong Anthropic Principle (in one of its many versions) *seems* to say that we might be able to prove the existence of God by virtue of an incredible statistical anomaly. The chances of the universe coming out badly were so high that only a deliberate intervention could have altered the odds in its favour.

In fact the Anthropic Principle doesn't actually promote this idea *per se*. The Principle is rather like Drake's Greenbank idea; it's a broad tool for discussion, not a refined predictive theory.

There is an alternative scenario within the Anthropic framework which disposes of the 'right-first-time' mystery.

THE INEVITABLE UNIVERSE

A certain class of star explodes towards the end of its life cycle. Matter is propelled inwards, as well as outwards. The star's old core is compressed. If this core subsequently drifts towards another star, or into a dust cloud, it drags in more material. The

AN ORDERED UNIVERSE?

core becomes so massive that it collapses under the pressure of its own gravity, until it becomes so tiny, it has no size at all. This 'singularity', infinitely dense, is called a 'Black Hole'.

Given time, a Black Hole could swallow a galaxy.

It has often been speculated that the 'other side' of a Black Hole consists of a *White* Hole in another universe, spewing out matter in a Big Bang. The general rules of physics break down when matter is sucked into a Black Hole, and they are created anew when matter is expelled from a Big Bang singularity. A Big Bang emerges from a dimensionless point of infinitely compressed matter. The most powerful Black Holes *are* single, dimensionless points of infinitely compressed matter. The two singularities, Black and Bang, are equivalent.

Surely a Black Hole would have to swallow an entire universe in one dimension, before it could spit it out into another? Not necessarily. Consider this: our cosmos has created many Black Hole singularities from old wreckage. We know this because we have found them. It follows that it doesn't require an entire universe to make one singularity. However, it requires only one singularity to make an entire universe ....

The idea that some Black Holes are capable of creating new universes is highly contentious, but it might explain how universes can chance upon success by a process similar to evolution. In November 1996, two biologists, John Maynard Smith of the University of Sussex, and Eörs Szathmary of the Collegium Budapest in Hungary, wrote an informal essay for the science journal *Nature*, entitled *On The Likelihood of Habitable Worlds*. They adapted ideas from a number of physical models.

This is how their theory works:

Our universe must be a good one, because we are here to discuss it. The universe's many Black Holes should be spewing out many new universes. We don't see them or interact with them. They are in another set of dimensions in time and space.

A universe can be considered 'weak' if its four fundamental forces are not well balanced. A weak universe is not so likely to create Black Holes. It won't generate other universes.

A universe like ours, good at making Black Holes and ideal for generating life, has no way of passing these strengths to any subsequent 'generations'. All the universe's prevailing laws of physics are scrambled by its own Black Holes. Weak universes may still be manufactured by most of these Holes. In this regard, universes cannot truly evolve. However, good universes with many Black Holes will tend to make many more universes. The more that are created, the higher the chances that some of *them* will be good. And so on. The process is exponential.

With this, the Anthropic dilemma is reversed. It is almost inevitable that we should find ourselves in a good universe.

There's one slight problem. Where did the first universe come from, good or poor? This question is phrased around our human understanding of time. Try using the word 'first' outside the context of time. Time is an emergent property of a universe and its physical laws, so our question is meaningless.

With arguments such as these, different interpretations of the Anthropic Principle can support opposite conclusions:

1) The suitability of this universe for the emergence of life is so unlikely, it must have come about through the intervention of a Divine Creator. Therefore it can be assumed that the universe has some higher meaning and purpose.

2) The suitability of our universe is inevitable. No divine intervention is required to explain our existence. The universe is without any particular meaning unless we choose to assign such a meaning ourselves.

CHAPTER TEN
THE LIMITS OF SCIENCE

SO FAR, EVERY discussion in this book has applied to things in the material realm that are subject to remorseless physical laws. Those who believe that the cosmos has some greater spiritual significance may have been depressed by the attention paid to atoms, molecules, evolution, and the other supposedly soulless constructs of the scientific viewpoint.

The Darwinian description of life has achieved a fantastic degree of detail and sophistication in recent times. Reductionist biologists regard the cosmos in general, and life in particular, as 'reducible' into its constituent molecular and atomic components. In a celebrated and controversial thesis, *The Selfish Gene* (1976), Richard Dawkins outlined with razor-sharp clarity how the DNA molecule creates organisms and controls every aspect of their behaviour. There has been no intervention whatsoever from a Divine Creator, he believes.

'The molecules of DNA are survival machines,' Dawkins states. 'They swarm in huge colonies, safe inside gigantic lumbering robots, sealed off from the outside world, communicating with it through tortuous indirect routes, manipulating it by remote control. They created us, in body and in mind. Their preservation is the ultimate rationale for our existence.'

It seems that all characteristics of an advanced life form can be traced to evolution, including what might seem the greatest miracle of all: consciousness, the very aspect of our humanity which impels us to investigate nature in the first place. If we take a moment to look at a *possible* reductionist explanation for consciousness, it might illuminate why we find the cosmos so deeply unknowable.

The best reductionist assessment of self-awareness can be found in Nicholas Humphrey's book, *Consciousness Regained* (1984). He believes that all our self-awareness, and that of other smart animals, derives from a particular evolutionary circumstance: we are social animals, living in groups. Interacting with others of our kind involves making bargains. You scratch my back, and I'll scratch yours. You share your food, and I'll let you share mine. Such cooperative interactions benefit individuals. Sometimes it's easier to survive in a group than alone.

A skilled cooperator gains a greater chance of living well and bringing up healthy children in a protected communal environment. Some of the burdens of life will be eased by the assistance of other individuals. He will benefit from energy savings. The cooperative *type* will prevail through future generations. The genes of aggressive loners stand less of a chance of being passed on, because everyone else will avoid them. They won't benefit from the generosity of others. Gradually, the uncooperative bloodline will be overtaken.

Nevertheless, a balance has to be struck. If an individual is generous all the time, others will take advantage. A few days' exploitation at the hands of slightly less generous companions could be damaging. Relationships between individuals have to be mainly cooperative but sometimes selfish, if cooperativeness as a whole is to work for groups and individuals alike.

Social animals have to read the faces and gestures of others to see what mood they're in. The cleverest individuals in any group can also work out how others are responding to *them*. They can manipulate the behaviour of other individuals to their own advantage. They develop a rough mental model of themselves, a kind of 'simulation', for predicting how others will respond when they approach.

THE LIMITS OF SCIENCE

In modern man, this interior model of self has become so useful and so persuasive that we regard it as real. We have assigned ourself 'selves'. Humphrey argues that this evolved to give individuals an advantage in social groups. It is not a 'soul'.

There are major tautologies in this line of reasoning, but it's a start. Humphrey also proposes that our religious impulses stem from our social desire to interact with the inanimate world around us, just as we interact with each other. 'Primitive (and not so primitive) people attempt to *bargain* with nature through prayer, through sacrifice or through ritual persuasion. In doing so, they are explicitly adopting a social model, expecting nature to participate in a transaction. But nature will not transact with men. She goes on her own way regardless.'

Our search for life in the cosmos, and alien intelligence in particular, may represent just another attempt to transact with inanimate nature.

The evolutionary origins of our intelligence present another difficulty for those of us who try to understand the cosmos. Our type of mind may not be suited for such a task.

WHY WE CANNOT KNOW THE TRUTH OF THINGS

Nature has adapted us to deal with the physical, sensory world. Other animals perceive their world differently. We cannot begin to imagine the thoughts that might form in the mind of a bat, 'seeing' the world by sound more than by sight. We cannot see flowers with an insect's compound eyes, except by using special cameras to translate the ultraviolet patterns which insects see into an optical blue that *we* can see. Likewise, the echo-located world of a dolphin is unknowable to us.

Different species perceive different realities. A species picked at random could not be expected to perceive the wide external reality of the cosmos as a whole. Why, then, should we expect to perceive it ourselves? So far, there has been no evolutionary pressure for any creature on earth to build any wider a picture of the surrounding world than is required for eating, breeding and eluding predators.

The scientific viewpoint, of necessity, is an invention of our animal minds. It may be that an animal mind is not capable of understanding anything beyond the animal circumstance. We human animals try to discover the mysteries of the universe with minds adapted for survival on earth, and therefore ill-adapted for speculations about space. The fact that we've been privileged to gain *any* insights into the deeper cosmos is remarkable.

Scientists working in the field of subatomic quantum physics often debate this issue. Their elusive branch of research reveals the limits of our ability to comprehend the nature of reality.

The fundamental units of matter do strange things. Electrons can go backwards in time, or appear in two places at once. Our disturbance of one electron may instantaneously alter the path of another very distant electron. A message seems to pass between them at speeds faster than light. When we probe an atom and its constituents, we can be sure that the values we observe (location and momentum) are not what they *would* have been had we left the atom alone. On the other hand, if we leave the atom alone, we find out nothing about it.

These are commonplaces of quantum physics, the most baffling and at the same time most successful and elaborate realm of human scientific endeavour. The cosmos is made entirely from atomic and subatomic entities which behave in a way that defies conventional human understanding.

We mentioned that a knowledge of the workings of a single atom would reveal the truths about the cosmos as a whole. We cannot know the atom. Therefore the cosmos, at its heart, remains mysterious. So does the basic mechanism of life.

In 1984, Peter Medawar, a respected biologist, wrote an essay entitled *The Limits of Science*. He warned that scientists were not entitled to say that they could explain everything in terms of scientific constructions—atoms, electrons, gravity fields, and so forth—which could not *themselves* be explained. With crisp logic, he argued that science can answer only scientific questions, and many questions are just not scientific.

'There are questions children ask. How did everything begin? What are we all here for? What is the point of life? It would be universally agreed that it is not to science that we should look for answers to those questions. There is, then, a limit to scientific understanding.'

THE LIMITS OF SCIENCE

The reductionists cannot claim all knowledge, Medawar says. 'It is not just to science, but to metaphysics, imagination and religion that we must turn for answers to questions that have to do with first and last things.'

LIFE INTO DUST AND DUST INTO LIFE

By weight, our bodies are 10 per cent hydrogen, 60 per cent oxygen and 20 per cent carbon. The last 10 per cent is taken up principally by nitrogen, calcium, phosphorous, sulphur, sodium and magnesium, iron and copper: all the elements that we have discussed at length in previous chapters.

The atoms which constitute these substances were created not in our sun but by other stars long since dead, as we have seen. They are not our atoms. They do not belong to us, nor to our bodies. They are just passing through.

Every moment of our lives, we exhale carbon dioxide. The atoms in that gas, we can be sure, will have existed 5000 million years ago, and more. Those atoms will continue to exist for billions of years after we are gone; in soil, in leaves, in dogs, in cats, in flowers, and in the air breathed by our descendants.

Our sun has burned for five billion years, and will burn for five billion more; after which, it will have used up a crucial proportion of its available hydrogen. At this point, things will take a different turn. The sun will become an unpredictable monster, prone to sudden nuclear shutdowns and re-ignitions, expansions and contractions, alternating over several millennia.

The sun will end its days as a red giant—so gigantic that its wayward mantle of gases will swallow up half the planets of the solar system, including the earth. Everything on our planet that lives and breathes will be utterly vaporized. The grass, the trees, even the toughened lichen that clings to granite will be blasted into atoms. The oceans will boil away to the last drop. The ice caps will vaporize. The sands on every beach in the world will fuse into glass. Mercury and Venus will vanish. If earth survives the storm, it will remain only as a scorched lump of rock.

The sun's final life-giving service will be to scatter all our atoms elsewhere, so that new solar systems and new creatures can come into being.

FURTHER READING

CLASSIC SPECULATIONS ABOUT MARS
Lowell, Percival. *Mars*
    Boston: Houghton Mifflin, 1896
Lowell, Percival. *Mars and its Canals*
    New York: Macmillan, 1906
Lowell, Percival. *Mars as the Abode of Life*
    New York: Macmillan, 1908
Wells, H.G. *The War of the Worlds*
    London: Heinemann, 1898

EARLY PLANS FOR MARS EXPLORATION
Braun, Wernher von. *The Mars Project* (English translation)
    Urbana, Illinois: University of Illinois Press, 1953
Braun, Wernher von. 'Can We Get to Mars?'
    Collier's, April 30, 1954
Ordway, Frederick and Liebermann, Randy. *Blueprint for Space*
    Washington DC: Smithsonian Institution Press, 1992
Sagan, Carl. 'Mars: A New World to Explore'
    National Geographic, Vol. 132, No. 6, Dec. 1967

MODERN SPACE HARDWARE & HISTORY
Gatland, Kenneth. *The Illustrated Encyclopedia of Space Technology*
    London: Salamander, 1989
Launius, Roger. *Nasa: A General History of the US Space Program*
    Malabar, Florida: Krieger, 1994
Powers, Robert M. *Planetary Encounters*
    London: Sidgwick & Jackson, 1982
Smith, Arthur. *Planetary Exploration: Three Decades of Planetary Probes*
    Wellingborough, England: Thorsons Group, 1988

FURTHER READING

THE MARINER IX MARS MISSION
Hartmann, W. *The New Mars: The Discoveries of Mariner IX*
    Washington DC: Nasa SP-337 US Govt. Printing Office, 1974
McCauley, J. and Hipsher, H. *Mars As Viewed by Mariner IX*
    Washington DC: Nasa SP-329 US Govt. Printing Office, 1974

THE VIKING PROJECT
Cooper, Henry S.F. 'The Search for Life on Mars'
    The New Yorker, Feb. 5, 1979, contd. Feb. 12, 1979
Gore, R. 'Mars As Viking Sees It'
    National Geographic, Jan. 1977
Levin, Gilbert V. *Viking Results, Chemistry or Biology?*
    Paper presented to the International Tesla Society Mars Forum,
    at Colorado Springs, on Nov. 13, 1993. Available from
    Biospherics Inc., 12051 Indian Creek Court, Beltsville,
    Maryland 20705, USA
Oyama, Vance, *et al.* 'Findings of the Viking GEX Experiment'
    Nature, Vol. 265, Jan. 13, 1977
Spitzer, Cary. *Viking Orbiter Views of Mars*
    Washington D.C: Nasa SP-441 US Govt. Printing Office, 1980
Viking Scientists. 'Viking I Mission Results'
    Science, Vol. 193, No. 4255, Aug. 1976
Viking Scientists. 'Viking I Mission Results' (second article)
    Science, Vol. 194, No. 4260, Oct. 1976
Viking Scientists. 'Viking II Mission Results'
    Science, Vol. 194, No. 4271, Dec. 1976
Viking Imaging Team. *The Martian Landscape*
    Washington D.C: Nasa SP-425 US Govt. Printing Office, 1978

MARTIAN METEORITE ALH84001
McKay, David, Gibson, Everett, *et al.* 'Search for Past Life on Mars:
Possible Relict Biogenic Activity in Martian Meteorite ALH84001'
    Science, Vol. 273, No. 5277, Aug. 16, 1996

FUTURE MARS EXPLORATION PROPOSALS
Stoker, Carol and Emmart, Carter. *Strategies for Mars*
    San Diego, California: Univelt, 1996
Zubrin, Robert and Wagner, Richard. *The Case for Mars*
    New York: Free Press, 1996

'TERRAFORMING' MARS FOR HUMAN HABITATION
Allaby, Michael and Lovelock, James. *The Greening of Mars*
    London: André Deutsch, 1984

FURTHER READING

Clarke, Arthur C. *The Snows of Olympus*
    London: Victor Gollancz, 1994
Fogg, Martyn and Birch, Paul. 'Terraforming Mars'
    Journal of the British Interplanetary Society, Vol. 45, No. 8, Aug. 1992
McKay, C., Toon, O. and Kasting, J. 'Making Mars Habitable'
    Nature, Vol. 352, No. 6335, Aug. 8, 1991

FICTIONAL ACCOUNTS OF MARS EXPLORATION
Baxter, Stephen. *Voyage*
    London: Voyager, 1996
Bova, Ben. *Mars*
    New York: Bantam, 1992
Robinson, Kim Stanley. *Red Mars*
    New York: Bantam, 1992
Robinson, Kim Stanley. *Green Mars*
    New York: Bantam, 1993
Robinson, Kim Stanley. *Blue Mars*
    New York: Bantam, 1996

MOLECULAR ORIGINS OF TERRESTRIAL LIFE
Cairns-Smith, Arthur. *Seven Clues to the Origins of Life*
    Cambridge: Cambridge University Press, 1985
Miller, Stanley. *The Origins of Life on Earth*
    Englewood Cliffs, New Jersey: Prentice-Hall, 1974

THE 'PANSPERMIA' THEORY
Hoyle, Fred and Wickramasinghe, Chandra. *Evolution from Space*
    London: J.M. Dent, 1981
Hoyle, Fred and Wickramasinghe, Chandra. *Our Place in the Cosmos*
    London: J.M. Dent, 1993
Wickramasinghe, Chandra. 'Where Life Begins?'
    New Scientist, April 21, 1977

LIFE IN THE COSMOS
Asimov, Isaac. *Extraterrestrial Civilizations*
    New York: Fawcett, 1979
Barrow, J. and Tipler, F. *The Anthropic Cosmological Principle*
    Oxford: Clarendon, 1986
Beckwith, S. and Sargent, A. 'Circumstellar Disks and the Search for Neighbouring Planetary Systems'
    Nature, Vol. 383, Sep. 12, 1996
Goldsmith, D. and Owen, T. *The Search for Life in the Universe*
    Menlo Park, California: Benjamin/Cummings, 1980

FURTHER READING

Maynard Smith, J. and Szathmary, E. 'Likelihood of Habitable Worlds'
Nature, Vol. 384, Nov. 14, 1996
Sagan, Carl. *Cosmos*
London: Macdonald, 1980

INTERSTELLAR TRAVEL
Bond, Alan and Martin, Anthony. *Project Daedalus: The Final Report on the British Interplanetary Society Starship Study*
London: The British Interplanetary Society, 1978
Clarke, Arthur C. *Rendezvous with Rama*
London: Victor Gollancz, 1973

SCIENTIFIC APPROACHES TO LIFE & ITS SIGNIFICANCE
Dawkins, Richard. *The Selfish Gene*
Oxford: Oxford University Press, 1976
Dawkins, Richard. *The Blind Watchmaker*
London: Longman, 1986
Humphrey, Nicholas. *Consciousness Regained*
Oxford: Oxford University Press, 1984
Diamond, Jared. *The Rise and Fall of the Third Chimpanzee*
London: Vintage, 1991
Ferguson, Kitty. *The Fire in the Equations*
London: Bantam, 1994
Medawar, Peter B. *The Limits of Science*
New York: Harper & Row, 1984
Sagan, Carl. *The Dragons of Eden: Speculations on the Evolution of Human Intelligence*
New York: Random House, 1977